Succeed

Eureka Math®
Grade K
Modules 1–3

Published by Great Minds®.

Copyright © 2018 Great Minds®.

Printed in the U.S.A.
This book may be purchased from the publisher at eureka-math.org.
CMP   10 9 8 7 6 5 4 3 2

ISBN 978-1-64054-096-5

GK-M1-M3-S-06.2018

Learn ⬧ Practice ⬧ Succeed

Eureka Math® student materials for *A Story of Units*® (K–5) are available in the *Learn, Practice, Succeed* trio. This series supports differentiation and remediation while keeping student materials organized and accessible. Educators will find that the *Learn, Practice,* and *Succeed* series also offers coherent—and therefore, more effective—resources for Response to Intervention (RTI), extra practice, and summer learning.

Learn

Eureka Math Learn serves as a student's in-class companion where they show their thinking, share what they know, and watch their knowledge build every day. *Learn* assembles the daily classwork—Application Problems, Exit Tickets, Problem Sets, templates—in an easily stored and navigated volume.

Practice

Each *Eureka Math* lesson begins with a series of energetic, joyous fluency activities, including those found in *Eureka Math Practice.* Students who are fluent in their math facts can master more material more deeply. With *Practice,* students build competence in newly acquired skills and reinforce previous learning in preparation for the next lesson.

Together, *Learn* and *Practice* provide all the print materials students will use for their core math instruction.

Succeed

Eureka Math Succeed enables students to work individually toward mastery. These additional problem sets align lesson by lesson with classroom instruction, making them ideal for use as homework or extra practice. Each problem set is accompanied by a Homework Helper, a set of worked examples that illustrate how to solve similar problems.

Teachers and tutors can use *Succeed* books from prior grade levels as curriculum-consistent tools for filling gaps in foundational knowledge. Students will thrive and progress more quickly as familiar models facilitate connections to their current grade-level content.

Students, families, and educators:

Thank you for being part of the *Eureka Math*® community, where we celebrate the joy, wonder, and thrill of mathematics.

Nothing beats the satisfaction of success—the more competent students become, the greater their motivation and engagement. The *Eureka Math Succeed* book provides the guidance and extra practice students need to shore up foundational knowledge and build mastery with new material.

What is in the Succeed *book?*

Eureka Math Succeed books deliver supported practice sets that parallel the lessons of *A Story of Units*®. Each *Succeed* lesson begins with a set of worked examples, called *Homework Helpers*, that illustrate the modeling and reasoning the curriculum uses to build understanding. Next, students receive scaffolded practice through a series of problems carefully sequenced to begin from a place of confidence and add incremental complexity.

How should Succeed *be used?*

The collection of *Succeed* books can be used as differentiated instruction, practice, homework, or intervention. When coupled with *Affirm*®, *Eureka Math*'s digital assessment system, *Succeed* lessons enable educators to give targeted practice and to assess student progress. *Succeed*'s perfect alignment with the mathematical models and language used across *A Story of Units* ensures that students feel the connections and relevance to their daily instruction, whether they are working on foundational skills or getting extra practice on the current topic.

Where can I learn more about Eureka Math *resources?*

The Great Minds® team is committed to supporting students, families, and educators with an ever-growing library of resources, available at eureka-math.org. The website also offers inspiring stories of success in the *Eureka Math* community. Share your insights and accomplishments with fellow users by becoming a *Eureka Math* Champion.

Best wishes for a year filled with Eureka moments!

Jill Diniz

Jill Diniz
Director of Mathematics
Great Minds

Contents

Module 1: Numbers to 10

Module 2: Two-Dimensional and Three-Dimensional Shapes

Module 3: Comparison of Length, Weight, Capacity, and Numbers to 10

Grade K
Module 1

Color the things that are exactly the same. Color them so that they look like each other.

> I didn't color the birds because they are not exactly the same. One is big, the other is small. Plus, they are not flying the same way.

> These trees are exactly the same. They are the same kind of tree, and they are the same size. I colored them so that they look like each other.

Lesson 1:   Analyze to find two objects that are *exactly the same* or *not exactly the same.*

3

© 2018 Great Minds®. eureka-math.org

Name _____   Date _____

Color the things that are exactly the same. Color them so that they look like each other.

EUREKA MATH

Lesson 1:   Analyze to find two objects that are *exactly the same* or *not exactly the same*.

© 2018 Great Minds®. eureka-math.org

5

Draw a line between two objects that match. Use your words. "These are the same, but this one is _____, and this one is _____."

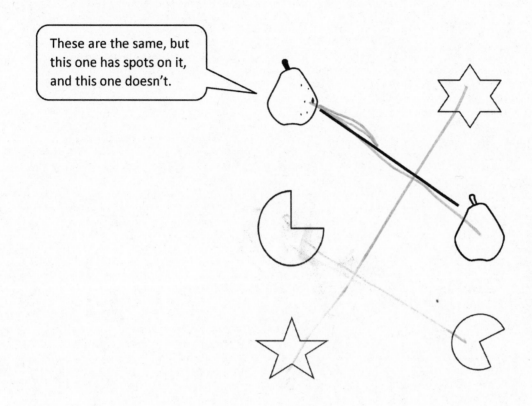

Name _____   Date _____

Draw a line between two objects that match. Use your words. "These are the same, but this one _____, and this one _____."

Make a picture of 2 things you use together. Tell why.

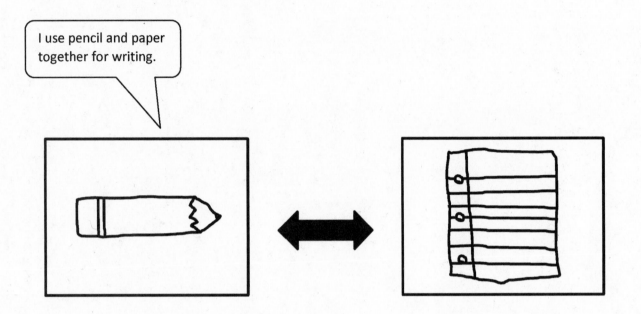

Name _____   Date _____

Draw something that you would use with each. Tell why.

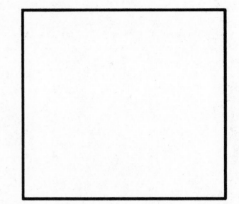

Make a picture of 2 things you use together. Tell why.

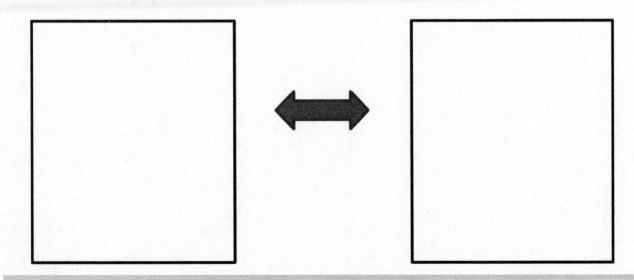

Make two groups. Circle the things that belong to one group. Underline the things that belong to the other group. Tell someone why the items in each group belong together. (There is more than one way to make groups!)

I sorted them into two groups: stuffed animals and real animals. How did you sort them?

Name _____   Date _____

Make two groups. Circle things that belong to one group. Underline the things that belong to the other group. Tell someone why the items in each group belong together. (There is more than one way to make groups!)

Use the cutouts. Glue the pictures to show where each belongs. Tell an adult how many are in each place.

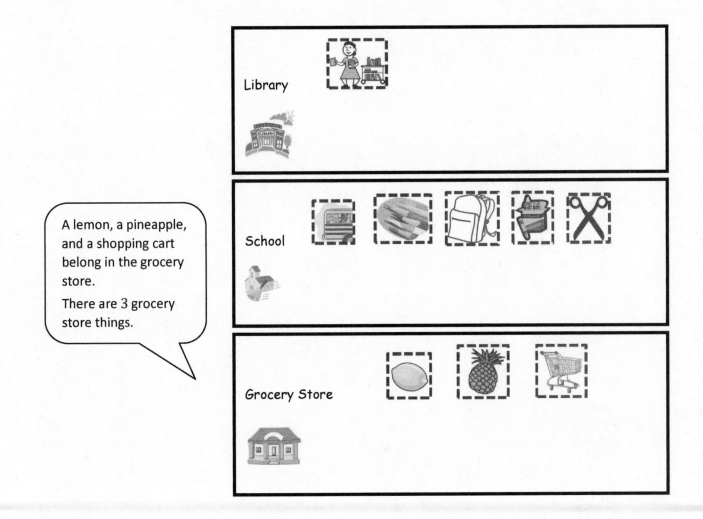

A lemon, a pineapple, and a shopping cart belong in the grocery store.

There are 3 grocery store things.

Lesson 5: Classify items into three categories, determine the count in each, and reason about how the last number named determines the total.

© 2018 Great Minds®. eureka-math.org

19

Name _____   Date _____

Use the cutouts. Glue the pictures to show where each belongs. Tell an adult how many are in each place.

Library

School

Grocery Store

EUREKA
MATH

Lesson 5:   Classify items into three categories, determine the count in each, and reason about how the last number named determines the total.

© 2018 Great Minds®. eureka-math.org

21

Homework Cutouts

Lesson 5:   Classify items into three categories, determine the count in each, and
reason about how the last number named determines the total.

23

Draw lines to put the treasures in the boxes.

I can sort by count!

Groups of 2 belong in the 2 box.

Groups of 3 belong in the 3 box.

Groups of 4 belong in the 4 box.

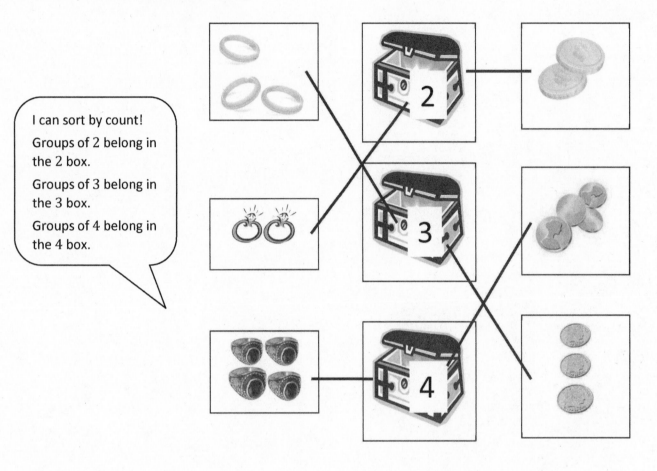

EUREKA
MATH®

Lesson 6:   Sort categories by count. Identify categories with 2, 3, and 4 within a
given scenario.

© 2018 Great Minds®. eureka-math.org

25

Name _____ Date _____

Draw lines to put the treasures in the boxes.

      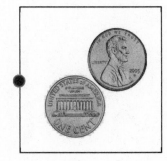

EUREKA MATH

Lesson 6:   Sort categories by count. Identify categories with 2, 3, and 4 within a given scenario.

27

© 2018 Great Minds®. eureka-math.org

Count and color.

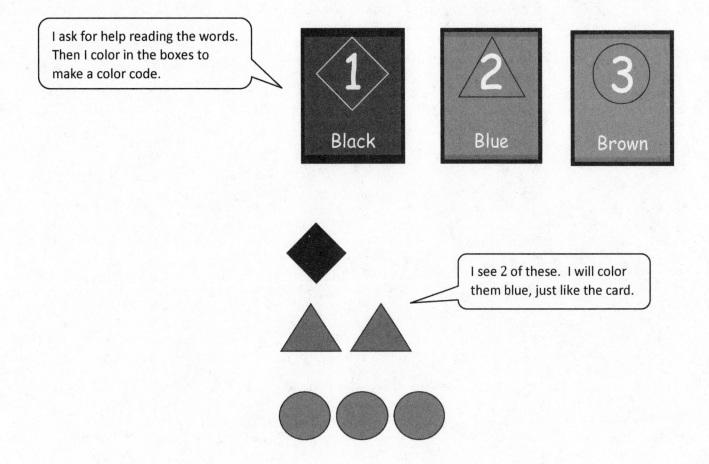

I ask for help reading the words. Then I color in the boxes to make a color code.

Black

Blue

Brown

I see 2 of these. I will color them blue, just like the card.

Lesson 7:   Sort by count in vertical columns and horizontal rows (linear configurations to 5). Match to numerals on cards.

29

Name _____   Date _____

Count and color.

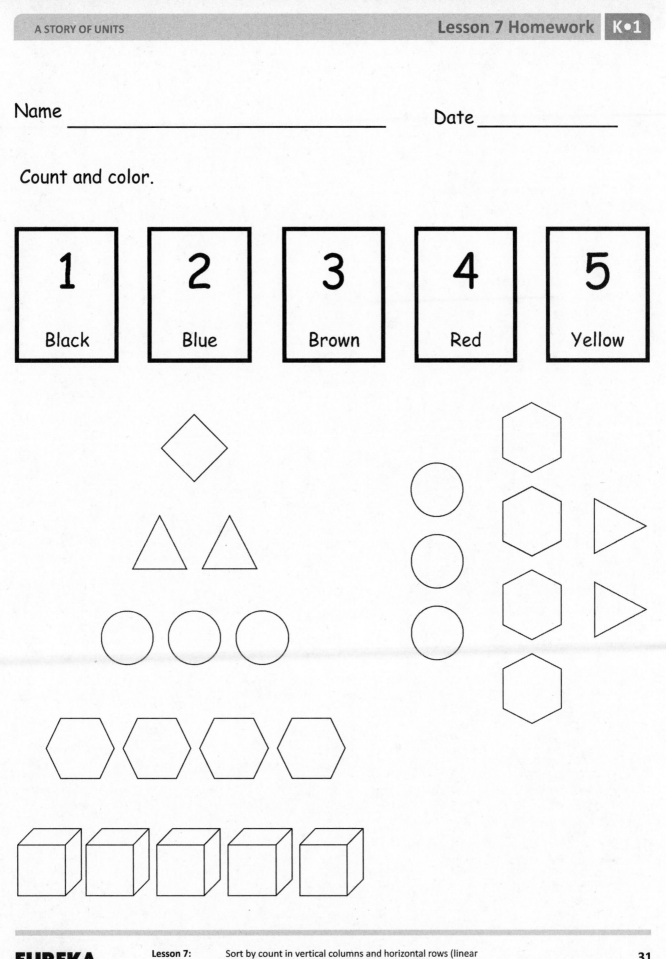

| 1 | 2 | 3 | 4 | 5 |
| Black | Blue | Brown | Red | Yellow |

EUREKA MATH®   Lesson 7:   Sort by count in vertical columns and horizontal rows (linear configurations to 5). Match to numerals on cards.   31

© 2018 Great Minds®. eureka-math.org

Count. Circle the number that tells *how many.*

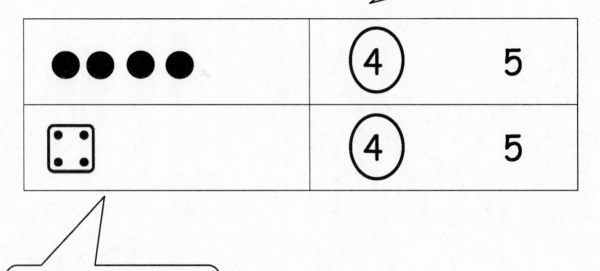

This one is easy! I counted 4 dots in a straight line. So I circle 4.

I counted 4 this time, too, but it looks different. I see 2 on the top and 2 on the bottom.

Lesson 8:   Answer *how many* questions to 5 in linear configurations (5-group), with 4 in an array configuration. Compare ways to count five fingers.

© 2018 Great Minds®. eureka-math.org

33

Name _____   Date _____

Count. Circle the number that tells how many dots in all.

● ● ● ●	4 5
• • • •	4 5
● ● ● ● ●	4 5
• • • • •	4 5
[die: 4]	4 5
[die: 4] [die: 1]	4 5
[die: 4]	4 5

EUREKA
MATH

Lesson 8:   Answer *how many* questions to 5 in linear configurations (5-group),
with 4 in an array configuration. Compare ways to count five fingers.

© 2018 Great Minds®. eureka-math.org

35

Count the circles, and box the correct number. Color in the same number of circles on the right as the shaded ones on the left to show hidden partners.

There are 4 circles: 3 of them are gray, and 1 is white. The hidden partners are 3 and 1.

3   [4]   5

I color in 3 circles.

I see 3 and 1 hiding inside of 4.

Lesson 9:   Within linear and array dot configurations of numbers 3, 4, and 5, find
hidden partners.

© 2018 Great Minds®. eureka-math.org

37

Name _____ Date _____

Count the circles, and box the correct number. Color the same number of circles on the right as the shaded ones on the left to show hidden partners.

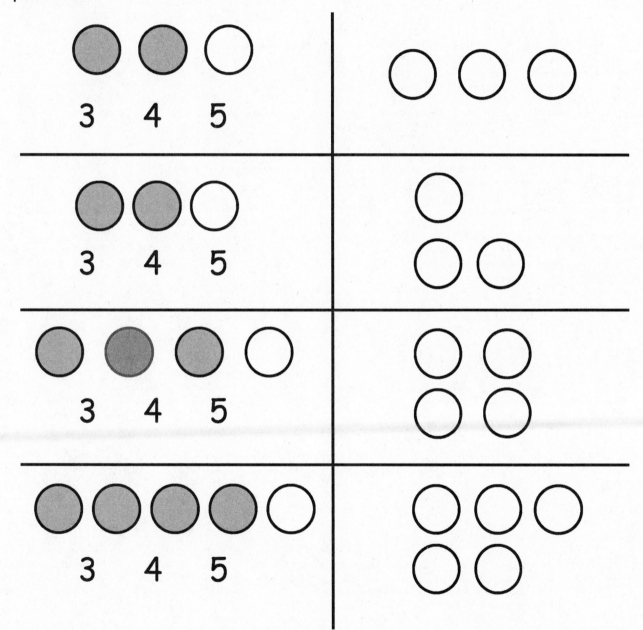

EUREKA MATH

Lesson 9: Within linear and array dot configurations of numbers 3, 4, and 5, find hidden partners.

© 2018 Great Minds®. eureka-math.org

Count how many. Draw a box around that number. Then, color 1 of the circles in each group.

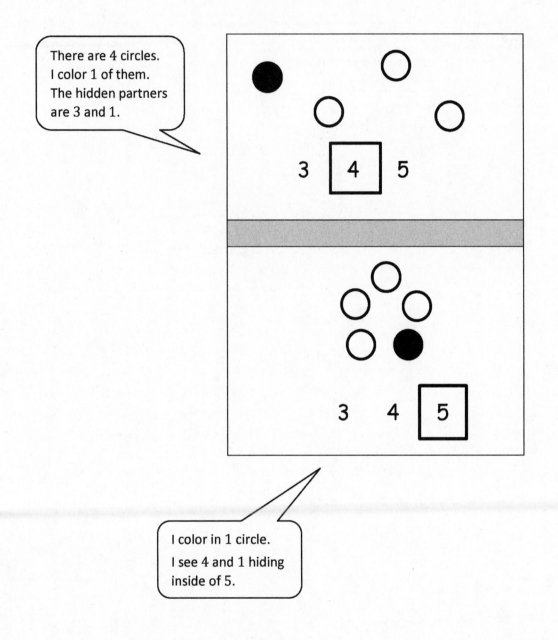

There are 4 circles.
I color 1 of them.
The hidden partners
are 3 and 1.

3   4   5

3   4   5

I color in 1 circle.
I see 4 and 1 hiding
inside of 5.

Lesson 10:   Within circular and scattered dot configurations of numbers 3, 4, and 5,
find *hidden partners*.

© 2018 Great Minds®. eureka-math.org

41

Name _____   Date _____

Count how many. Draw a box around that number. Then, color 1 of the circles in each group.

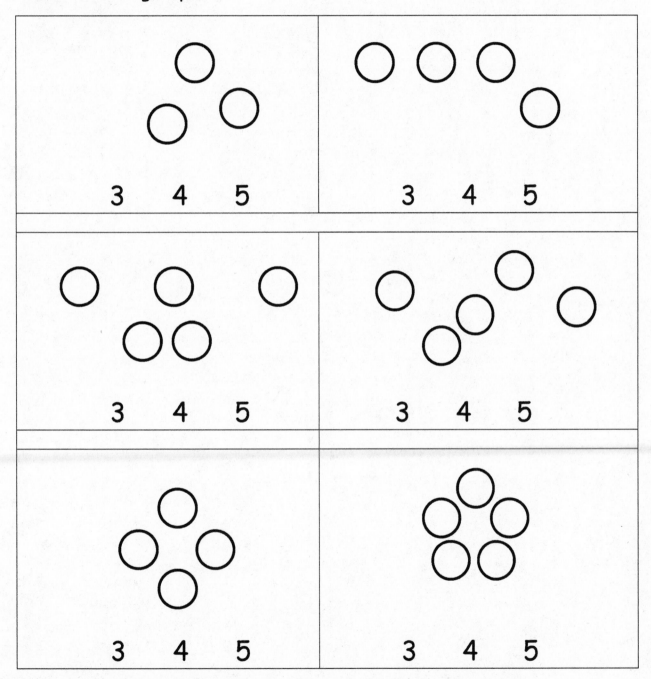

Talk to an adult at home about the hidden partners you found.

Lesson 10:   Within circular and scattered dot configurations of numbers 3, 4, and 5, find *hidden partners*.

© 2018 Great Minds®. eureka-math.org

43

Color the shapes to show 1 + 2. Use your 2 favorite colors.

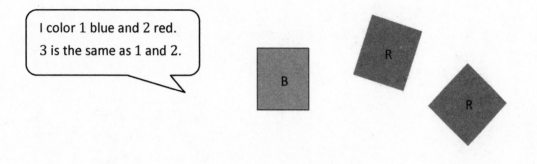

I color 1 blue and 2 red.
3 is the same as 1 and 2.

How many shapes are there?

Circle the number.   1   2   3   4   5

Lesson 11:   Model decompositions of 3 with materials, drawings, and expressions.
Represent the decomposition as 1 + 2 and 2 + 1.

45

© 2018 Great Minds®. eureka-math.org

Name _____   Date _____

Feed the puppies! Here are 3 bones. Draw lines to show 2 + 1.

Color the shapes to show 1 + 4. Use your 2 favorite colors.

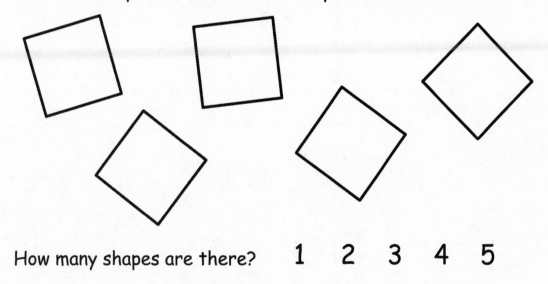

How many shapes are there?   1   2   3   4   5

Lesson 11:   Model decompositions of 3 with materials, drawings, and expressions.
Represent the decomposition as 1 + 2 and 2 + 1.

47

How many? Draw a line between each picture and its number.

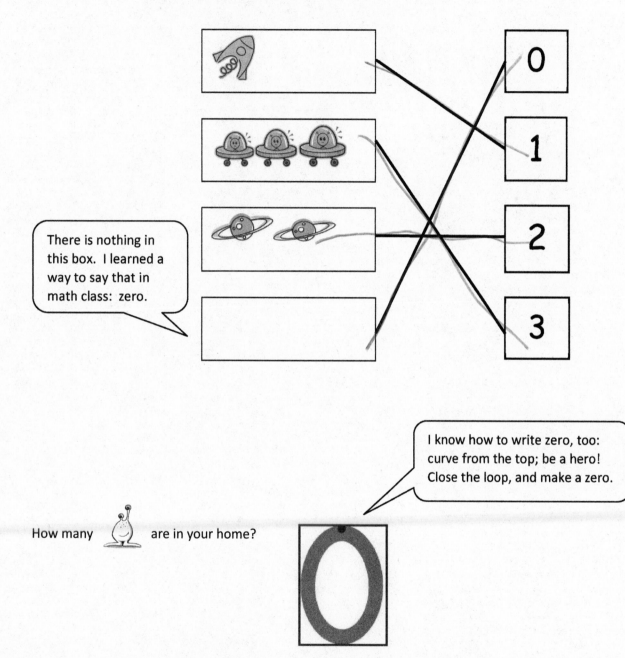

There is nothing in this box. I learned a way to say that in math class: zero.

I know how to write zero, too: curve from the top; be a hero! Close the loop, and make a zero.

How many are in your home?

Lesson 12: Understand the meaning of zero. Write the numeral 0.

49

EUREKA MATH

Name _____   Date _____

How many? Draw a line between each picture and its number.

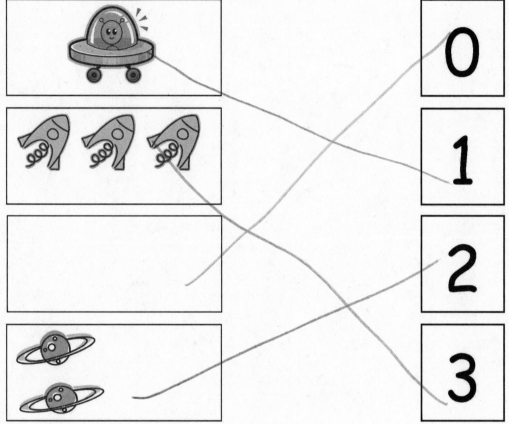

Write the number in the blank.

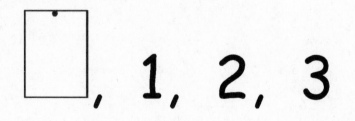

$\boxed{}$, 1, 2, 3

Lesson 12:   Understand the meaning of zero. Write the numeral 0.   51

© 2018 Great Minds®. eureka-math.org

Count the objects. Write how many.

1, 2.

I count 2 cats.
I write the
number 2.

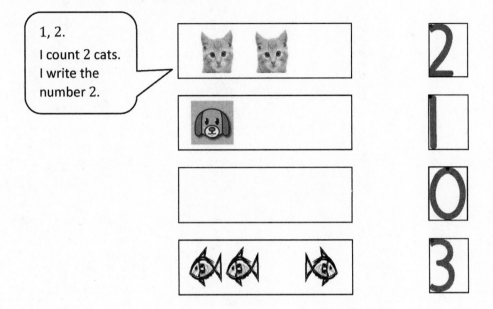

Write the missing numbers.

1, 2, 3   3, 2, 1, 0

Name _____   Date _____

Draw [●●▢▢▢] (two) pots.

How many?

Draw [●▢▢▢▢] (one) friend.

How many?

Draw [●●●▢▢] (three) toys.

How many?

How many pet monkeys 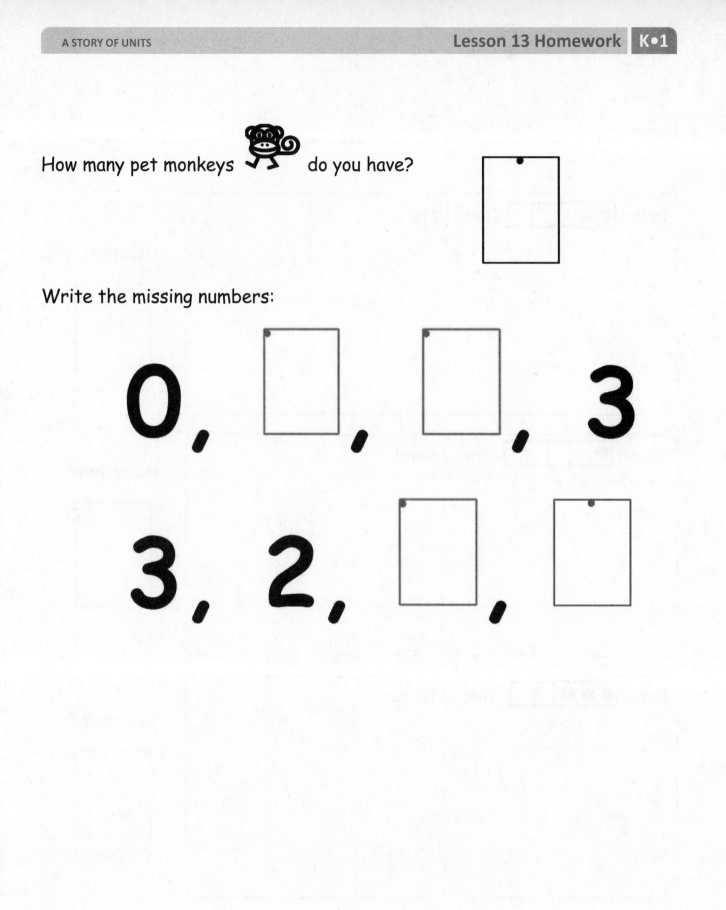 do you have?

Write the missing numbers:

0, ☐, ☐, 3

3, 2, ☐, ☐

EUREKA
MATH

Color the stars so that 1 is yellow and 2 are red.

I count 3 things. I color 1 star yellow and 2 stars red. When I take apart 3, its parts are 2 and 1.

There are stars.

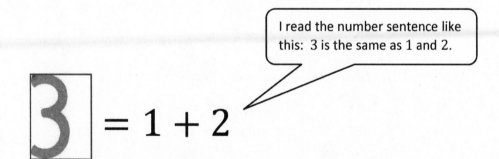

I read the number sentence like this: 3 is the same as 1 and 2.

$$3 = 1 + 2$$

EUREKA MATH®

Lesson 14:   Write numerals 1–3. Represent decompositions with materials, drawings, and equations, 3 = 2 + 1 and 3 = 1 + 2.

© 2018 Great Minds®. eureka-math.org

57

Name _____   Date _____

Color the shirts so that 1 is red and 2 are green.

There are $\boxed{3}$ shirts.

$\boxed{3}$ $= 1 + 2$

Color the balls so that 2 are yellow and 1 is blue.

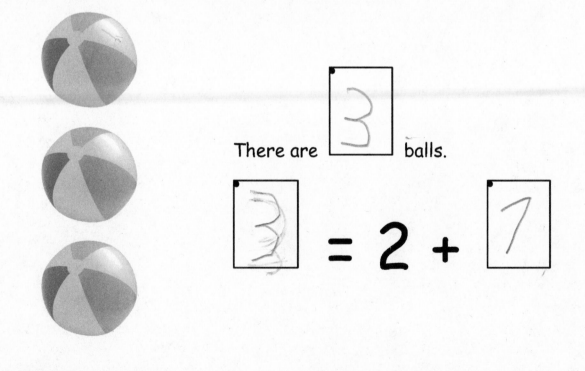

There are $\boxed{3}$ balls.

$\boxed{3}$ $= 2 +$ $\boxed{1}$

Lesson 14:   Write numerals 1–3. Represent decompositions with materials,
drawings, and equations, 3 = 2 + 1 and 3 = 1 + 2.

© 2018 Great Minds®. eureka-math.org

59

Count the shapes and write the numbers. Mark each shape as you count.

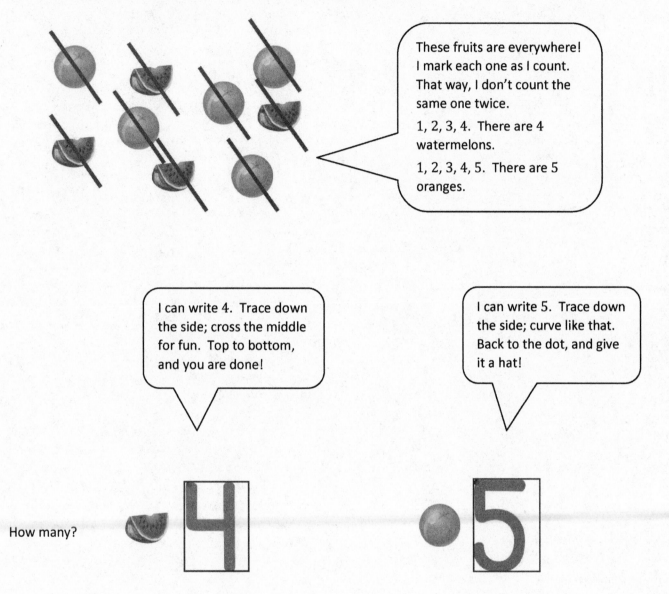

These fruits are everywhere! I mark each one as I count. That way, I don't count the same one twice.

1, 2, 3, 4. There are 4 watermelons.

1, 2, 3, 4, 5. There are 5 oranges.

I can write 4. Trace down the side; cross the middle for fun. Top to bottom, and you are done!

I can write 5. Trace down the side; curve like that. Back to the dot, and give it a hat!

How many?

EUREKA MATH

Lesson 15: Order and write numerals 4 and 5 to answer *how many* questions in categories; sort by count.

© 2018 Great Minds®. eureka-math.org

61

Name _____   Date _____

Count the shapes and write the numbers. Mark each shape as you count.

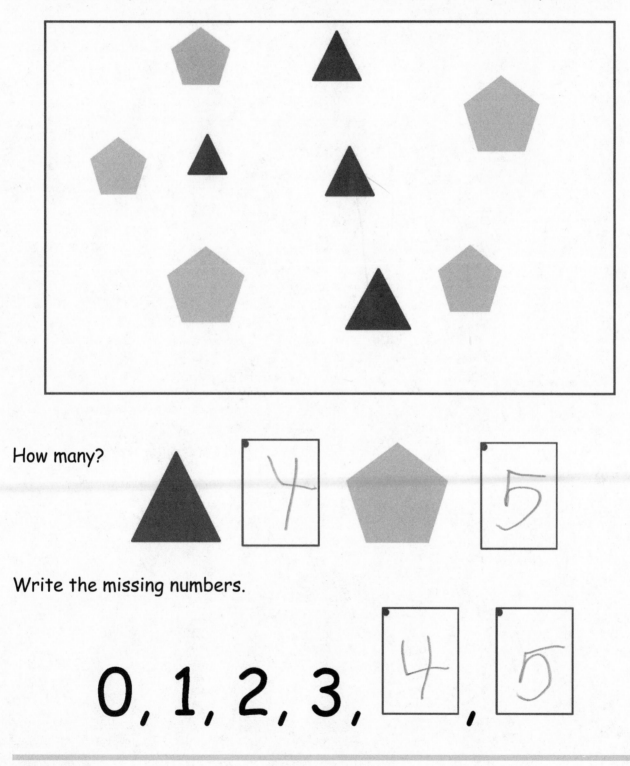

How many?

Write the missing numbers.

0, 1, 2, 3, 4, 5

Lesson 15:   Order and write numerals 4 and 5 to answer *how many* questions in categories; sort by count.

© 2018 Great Minds®. eureka-math.org

63

Write the missing numbers:

1, 2, 3, 4, 5

I can count up and down.

Counting out loud helps me find the missing number.

4, 3, 2, 1, 0

Draw 3 yellow fish and 2 green fish.

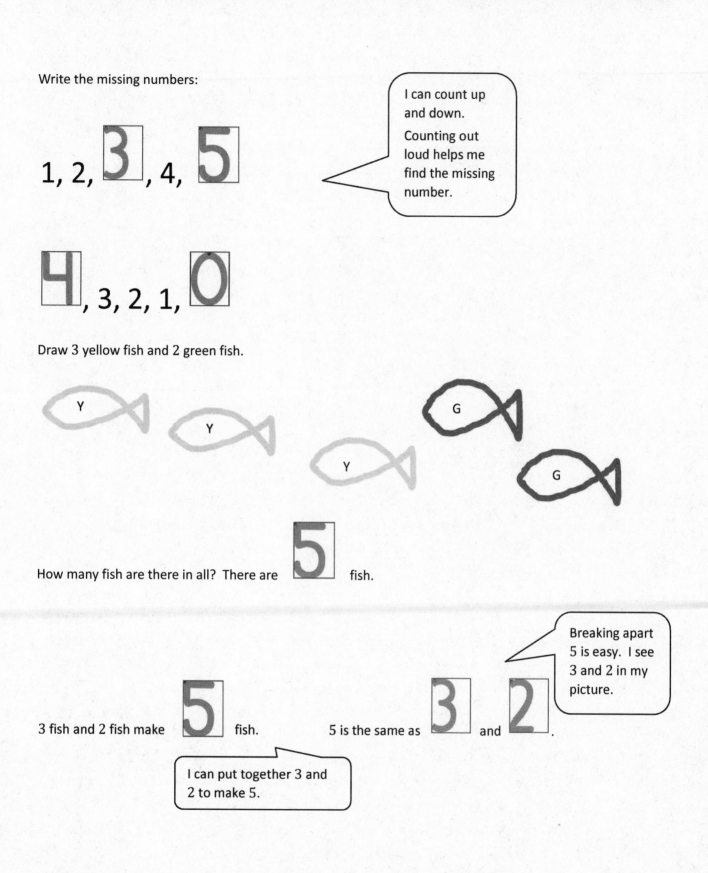

How many fish are there in all? There are 5 fish.

3 fish and 2 fish make 5 fish.   5 is the same as 3 and 2.

Breaking apart 5 is easy. I see 3 and 2 in my picture.

I can put together 3 and 2 to make 5.

EUREKA MATH

Lesson 16:   Write numerals 1–5 in order. Answer and make drawings of decompositions with totals of 4 and 5 without equations.

65

© 2018 Great Minds®. eureka-math.org

Name _____   Date _____

Write the missing numbers:

1, 2, 3 , 4, 5

5, 4 , 3, 2, 1

4 , 3, 2, 1, 0

0 , 1, 2, 3 , 4

Lesson 16:   Write numerals 1–5 in order. Answer and make drawings of
decompositions with totals of 4 and 5 without equations.

67

© 2018 Great Minds®. eureka-math.org

Draw 3 red fish and 1 green fish.

How many fish are there in all?

3 fish and 1 fish make ☐ fish.

Draw 2 happy faces and 3 sad faces.

How many faces are there in all? ☐

5 is the same as ☐ and ☐

Lesson 16:   Write numerals 1–5 in order. Answer and make drawings of decompositions with totals of 4 and 5 without equations.

EUREKA
MATH

Color 4.

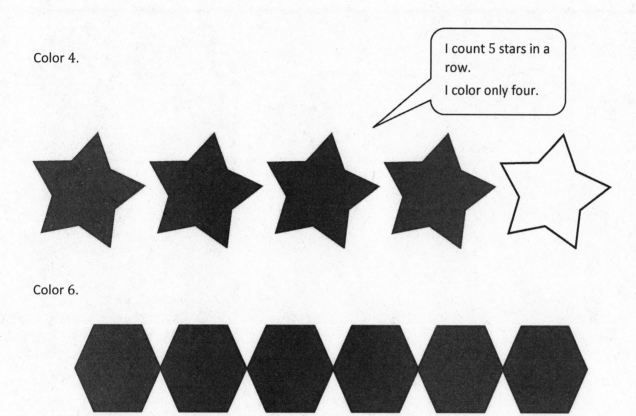

I count 5 stars in a row.
I color only four.

Color 6.

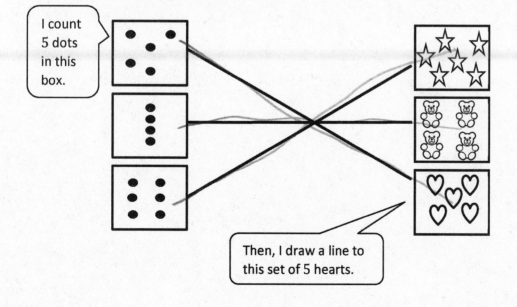

Connect the boxes with the same number.

I count 5 dots in this box.

Then, I draw a line to this set of 5 hearts.

Lesson 17: Count 4–6 objects in vertical and horizontal linear configurations and array configurations. Match 6 objects to the numeral 6.

© 2018 Great Minds®. eureka-math.org

69

Name _____   Date _____

Color 4.

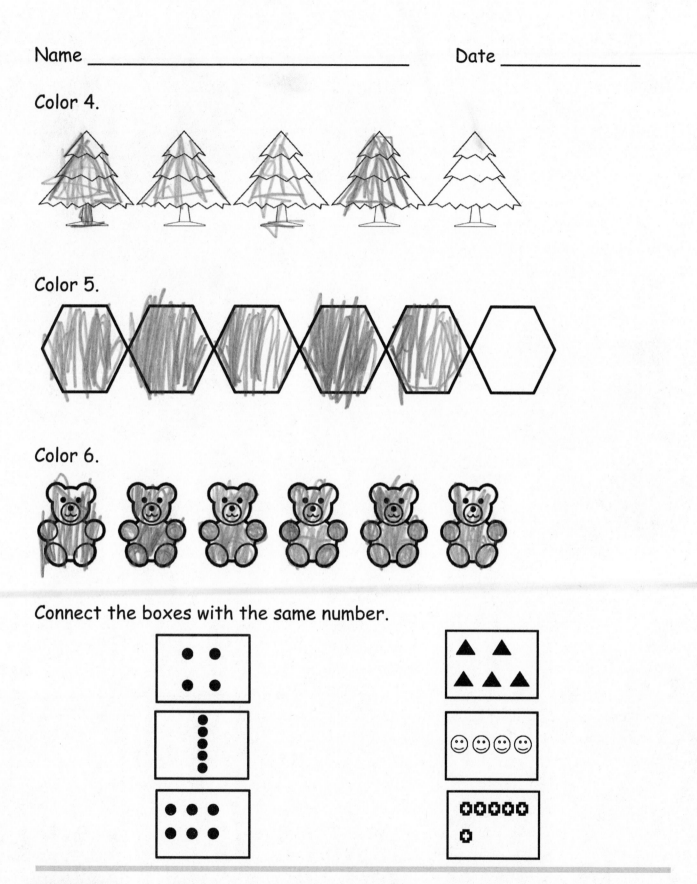

Color 5.

Color 6.

Connect the boxes with the same number.

EUREKA
MATH®

Lesson 17:   Count 4–6 objects in vertical and horizontal linear configurations and
array configurations. Match 6 objects to the numeral 6.

© 2018 Great Minds®. eureka-math.org

71

Color 4.

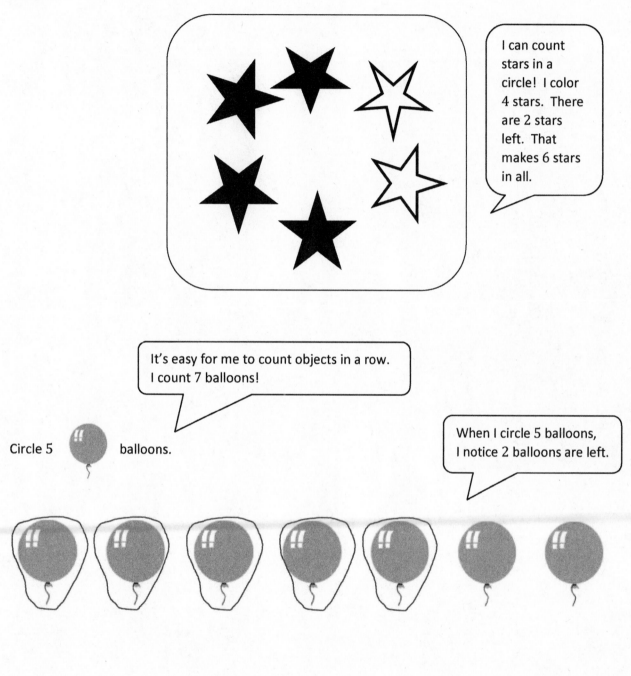

I can count stars in a circle! I color 4 stars. There are 2 stars left. That makes 6 stars in all.

It's easy for me to count objects in a row. I count 7 balloons!

Circle 5   balloons.

When I circle 5 balloons, I notice 2 balloons are left.

    Lesson 18:   Count 4–6 objects in circular and scattered configurations. Count 6 items out of a larger set. Write numerals 1–6 in order.   **73**

© 2018 Great Minds®. eureka-math.org

Name _____ Date _____

Color 5.

Color 6.

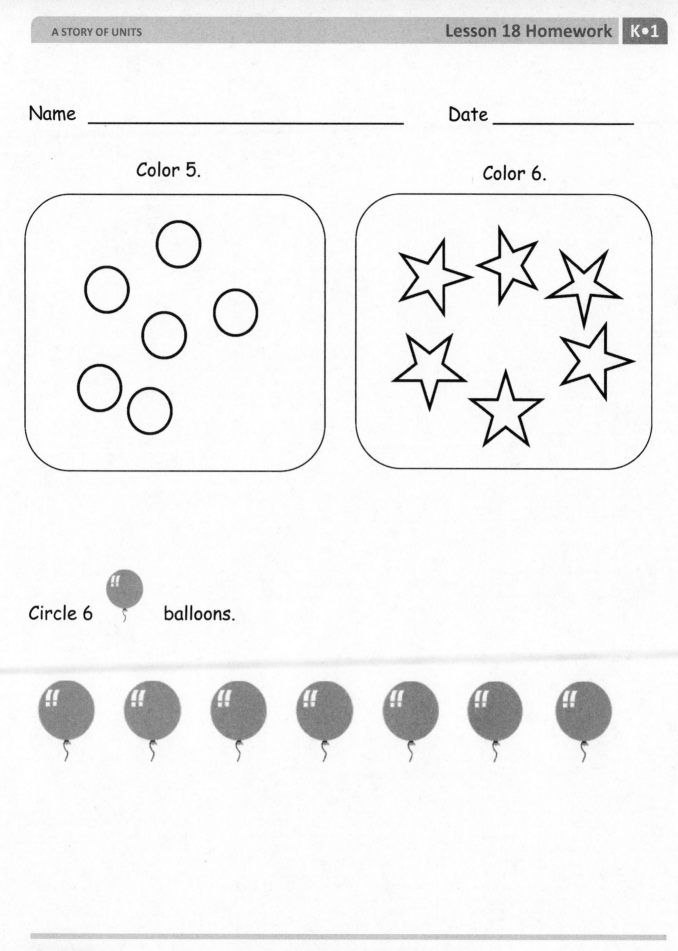

Circle 6 balloons.

Lesson 18: Count 4–6 objects in circular and scattered configurations.
Count 6 items out of a larger set. Write numerals 1–6 in order.

75

EUREKA
MATH®

© 2018 Great Minds®. eureka-math.org

• • • • •

5-group

Like fingers on a hand, we can make groups of 5 (and some more).

Draw a line from the numeral to the 5-group it matches.

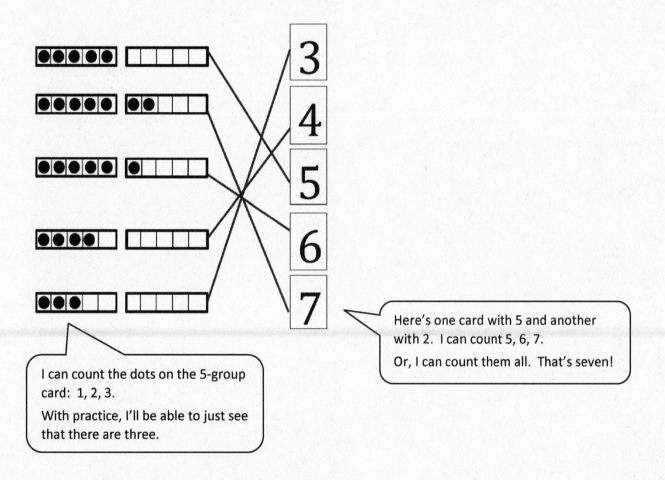

I can count the dots on the 5-group card: 1, 2, 3.

With practice, I'll be able to just see that there are three.

Here's one card with 5 and another with 2. I can count 5, 6, 7.

Or, I can count them all. That's seven!

Lesson 19:   Count 5–7 linking cubes in linear configurations. Match with numeral 7.
Count on fingers from 1 to 7, and connect to 5-group images.

77

© 2018 Great Minds®. eureka-math.org

Fill in the missing numbers.

I count up to 7, starting from any number.
Look at me! I can write my numbers!

4,5, 6, 7

7, 6, **5** , 4, **3** , 2

1, **2** , 3, **4** ,5, 6, 7

Lesson 19: Count 5–7 linking cubes in linear configurations. Match with numeral 7.
Count on fingers from 1 to 7, and connect to 5-group images.

Name _____   Date _____

Draw a line from the 5-groups to the matching numbers.

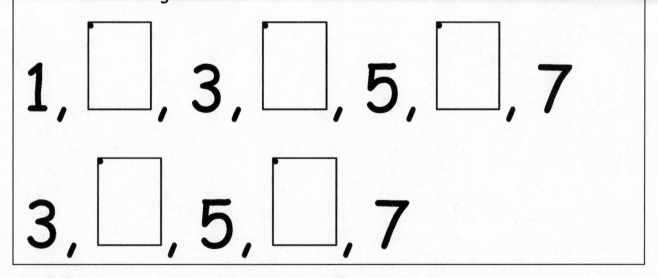

Fill in the missing numbers.

1, [ ] , 3, [ ] , 5, [ ] , 7

3, [ ] , 5, [ ] , 7

Lesson 19:   Count 5–7 linking cubes in linear configurations. Match with numeral 7.
Count on fingers from 1 to 7, and connect to 5-group images.

© 2018 Great Minds®. eureka-math.org

79

How many? Write the number in the box.

7

Look! I see 5 and 2 more! That makes 7.

I can count the triangles! Here is my counting path.

What's yours?

Count how many. Write the number in the box.

Draw a line to show how you counted the triangles.

7

There are 7 in all! "A straight line and down from heaven; that's the way we make a 7."

Lesson 20:   Reason about sets of 7 varied objects in circular and scattered configurations. Find a path through the scattered configuration. Write numeral 7. Ask, "How is your seven different from mine?"

© 2018 Great Minds®. eureka-math.org

81

Name _____ Date _____

How many? Write the number in the box.

Lesson 20:   Reason about sets of 7 varied objects in circular and scattered
configurations. Find a path through the scattered configuration.
Write numeral 7. Ask, "How is your seven different from mine?"

© 2018 Great Minds®. eureka-math.org

Count how many. Write the number in the box.
Draw a line to show how you counted the suns.

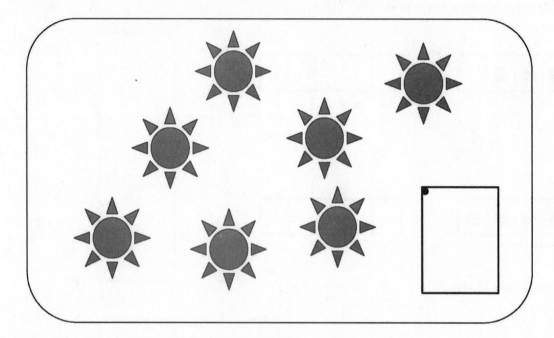

Count how many. Write the number in the box.
Draw a line to show how you counted the circles.

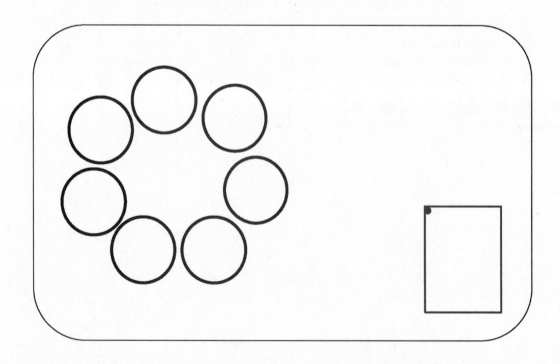

Lesson 20:   Reason about sets of 7 varied objects in circular and scattered
                        configurations. Find a path through the scattered configuration.
                        Write numeral 7. Ask, "How is your seven different from mine?"

© 2018 Great Minds®. eureka-math.org

EUREKA
MATH

Color 4 ladybugs red. Color 4 ladybugs yellow.

Count and circle how many.

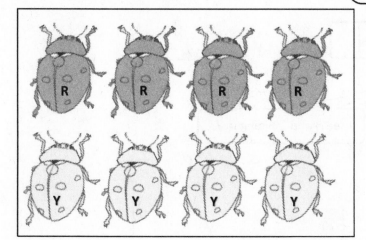

These two rows have the same number of ladybugs. I can see 4 and 4 hiding in 8.

6   7   ⑧

Color 4 ladybugs blue. Color 4 ladybugs orange.

Count and circle how many.

It doesn't matter whether the ladybugs are arranged in rows or columns; there are still 8 ladybugs in all!

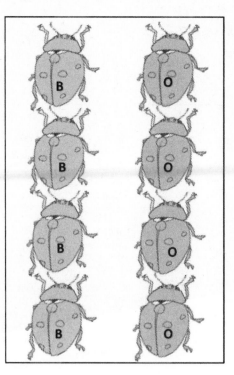

6   7   ⑧

Count how many. Write the number in the box.

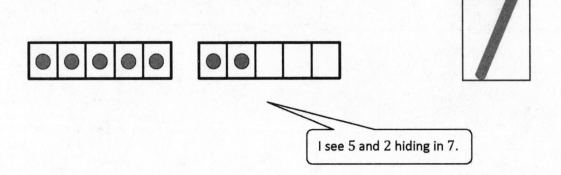

I see 5 and 2 hiding in 7.

© 2018 Great Minds®. eureka-math.org

Lesson 21: Compare counts of 8 in linear and array configurations. Match with numeral 8.

Name _____   Date _____

Color 4 squares blue.   Color 4 squares yellow.

Count and circle how many.

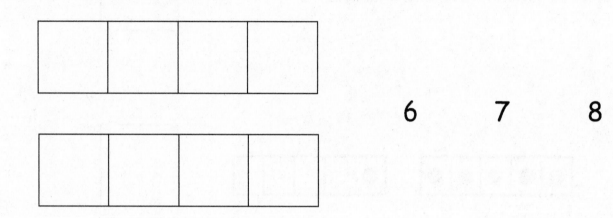

6   7   8

Color 4 squares green.   Color 4 squares brown.

Count and circle how many.

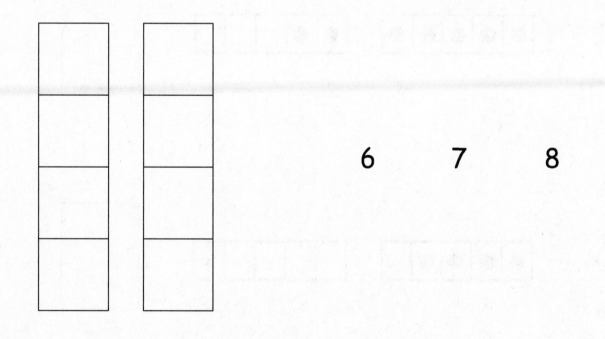

6   7   8

Lesson 21:   Compare counts of 8 in linear and array configurations. Match with
numeral 8.

© 2018 Great Minds®. eureka-math.org

87

Count how many. Write the number in the box.

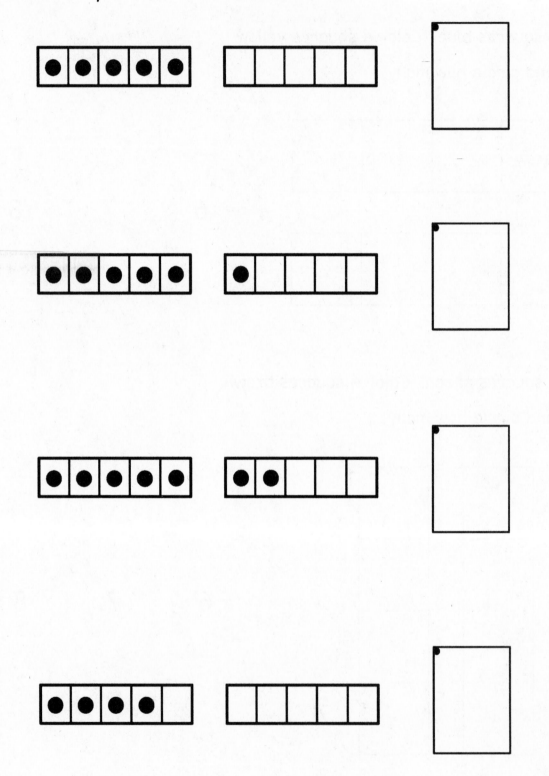

© 2018 Great Minds®. eureka-math.org

Draw 8 beads around the circle.

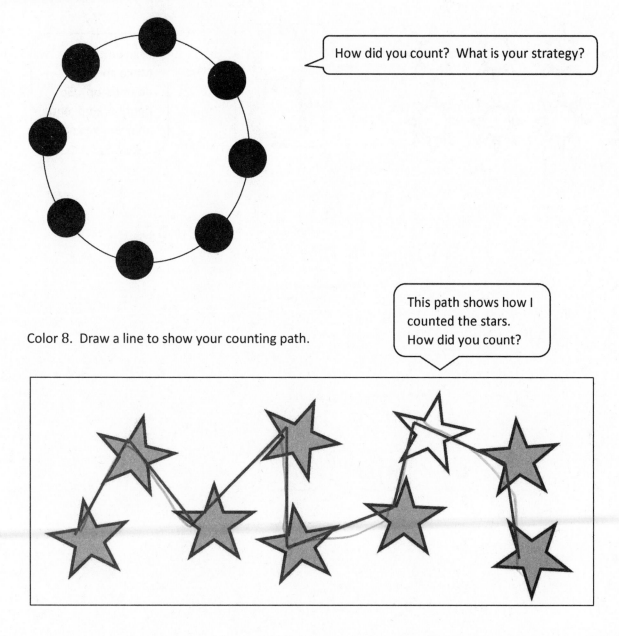

How did you count? What is your strategy?

Color 8. Draw a line to show your counting path.

This path shows how I counted the stars. How did you count?

Count how many. Write the number in the box

I can write 8. Make an S, and do not stop. Go right back up, and an 8 you've got!

Lesson 22: Arrange and strategize to count 8 beans in circular (around a cup) and scattered configurations. Write numeral 8. Find a path through the scattered set, and compare paths with a partner.

Name _____   Date _____

Draw 8 beads around the circle.

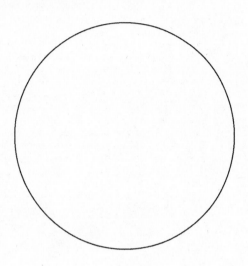

Color 8. Draw a line to show your counting path.

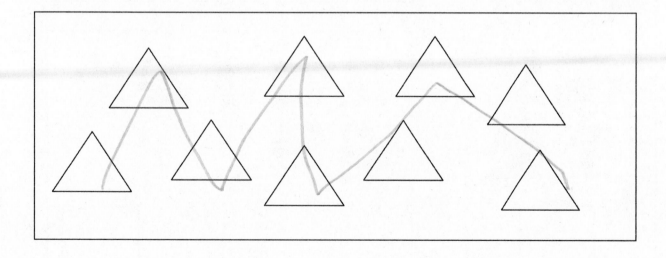

Lesson 22:   Arrange and strategize to count 8 beans in circular (around a cup) and
scattered configurations. Write numeral 8. Find a path through the
scattered set, and compare paths with a partner.

91

EUREKA
MATH

Count how many. Write the number in the writing rectangle.

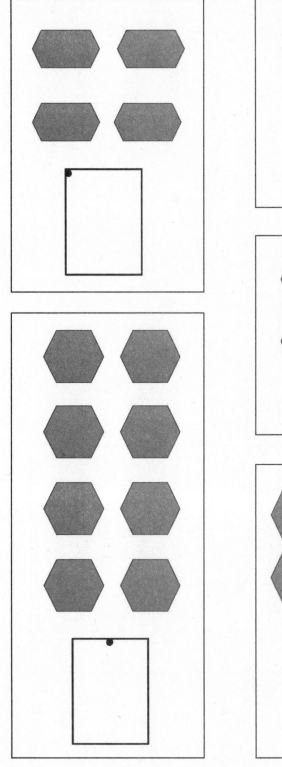

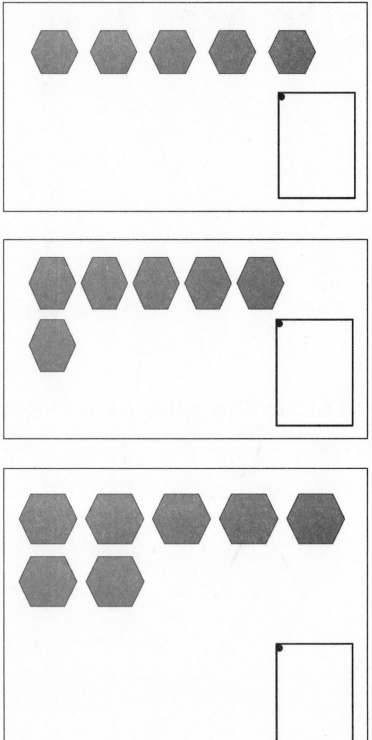

Lesson 22: Arrange and strategize to count 8 beans in circular (around a cup) and scattered configurations. Write numeral 8. Find a path through the scattered set, and compare paths with a partner.

Color 9 circles.

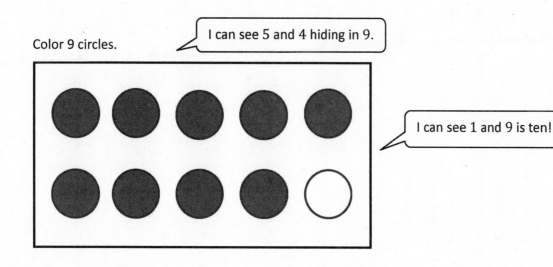

I can see 5 and 4 hiding in 9.

I can see 1 and 9 is ten!

Draw 9 shapes.

Do your shapes look like mine? There are so many ways to draw and arrange nine!

Name _____ Date _____

Color 9 shapes.

Color 9 shapes.

Draw 9 circles.

Draw 9 shapes.

Color 9 circles.

Look at me! I can count 9 circles scattered about. I don't count any circles more than once. I have a strategy. Do you?

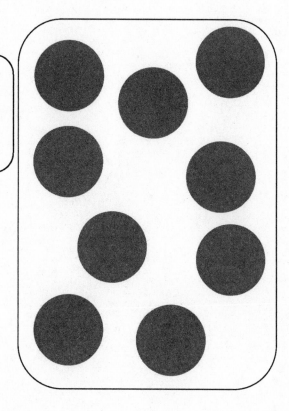

Count. Write the number in the box.

I remember how to write 9. A hoop and a line. That's the way we make nine!

EUREKA MATH

Lesson 24: Strategize to count 9 objects in circular (around a paper plate) and scattered configurations printed on paper. Write numeral 9. Represent a path through the scatter count with each object.

© 2018 Great Minds®. eureka-math.org

97

Name _____ Date _____

Color 9 circles.

Color 9 circles.

Draw 9 beads on the bracelet.

Count. Write the number in the box.

Color 5 suns. Color 5 more suns a different color.

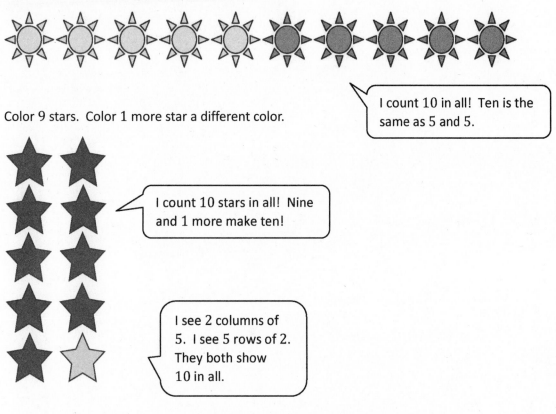

I count 10 in all! Ten is the same as 5 and 5.

Color 9 stars. Color 1 more star a different color.

I count 10 stars in all! Nine and 1 more make ten!

I see 2 columns of 5. I see 5 rows of 2. They both show 10 in all.

Draw 5 circles under the row of circles. Color 5 circles yellow. Color 5 circles green.

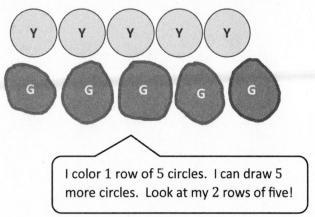

I color 1 row of 5 circles. I can draw 5 more circles. Look at my 2 rows of five!

Lesson 25: Count 10 objects in linear and array configurations (2 fives). Match with numeral 10. Place on the 5-group mat. Dialogue about 9 and 10. Write numeral 10.

© 2018 Great Minds®. eureka-math.org

101

Name _____   Date _____

Color 9 squares. Color 1 more square a different color.

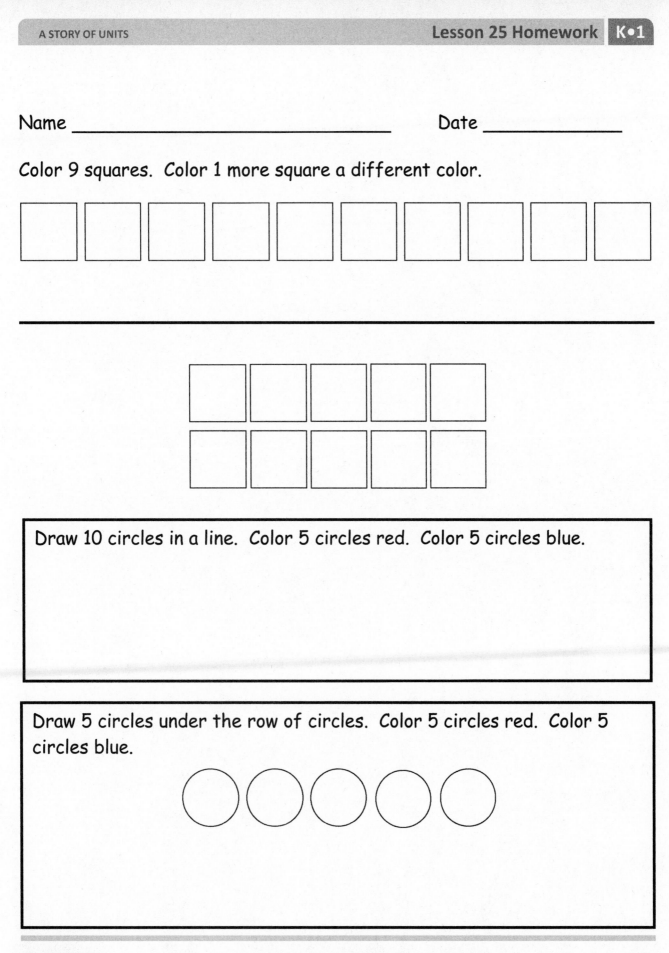

Draw 10 circles in a line. Color 5 circles red. Color 5 circles blue.

Draw 5 circles under the row of circles. Color 5 circles red. Color 5 circles blue.

Lesson 25:   Count 10 objects in linear and array configurations (2 fives). Match
with numeral 10. Place on the 5-group mat. Dialogue about
9 and 10. Write numeral 10.

103

EUREKA
MATH®

© 2018 Great Minds®. eureka-math.org

Draw 5 circles in a row. Draw another 5 circles in a row under them.

How many circles did you draw?

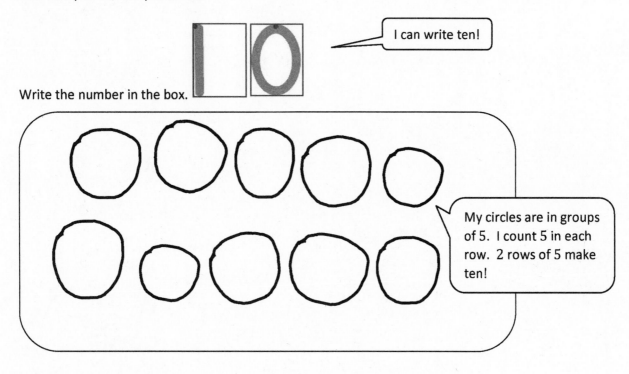

I can write ten!

Write the number in the box.

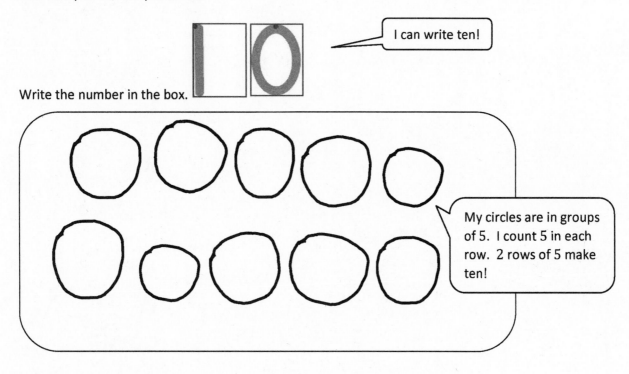

My circles are in groups of 5. I count 5 in each row. 2 rows of 5 make ten!

Write how many in the box.

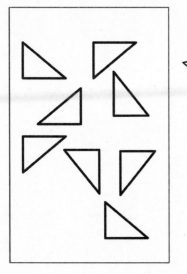

These triangles are not arranged in a line. But, I can count them all without counting twice. I've got a strategy!

Name _____ Date _____

Draw 5 triangles in a row. Draw another 5 triangles in a row under them.
How many triangles did you draw? Write the number.

Write how many.

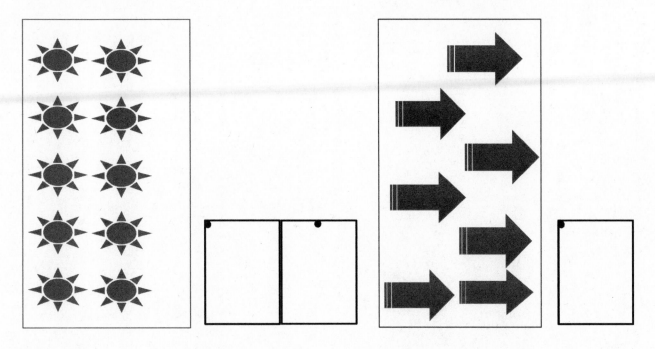

Write how many.

EUREKA MATH

Lesson 26: Count 10 objects in linear and array configurations (2 fives). Match
with numeral 10. Place on the 5-group mat. Dialogue about 9 and 10.
Write numeral 10.

© 2018 Great Minds®. eureka-math.org

107

Draw enough 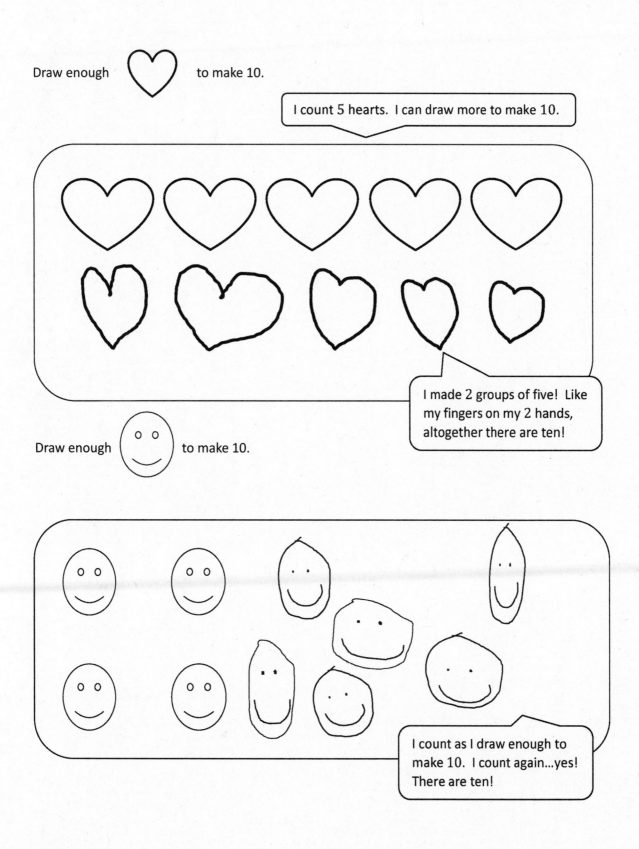 to make 10.

I count 5 hearts. I can draw more to make 10.

I made 2 groups of five! Like my fingers on my 2 hands, altogether there are ten!

Draw enough to make 10.

I count as I draw enough to make 10. I count again...yes! There are ten!

EUREKA MATH

Name _____   Date _____

Draw more 🌥 to make 10.

Draw more 🙂 to make 10.

EUREKA MATH

Lesson 27:   Count 10 objects, and move between all configurations.

© 2018 Great Minds®. eureka-math.org

111

Make up a story about 10 things in your house. Draw a picture to go with your story. Be ready to share your story at school tomorrow.

I remember math stories we acted out in class today. Stories like, "8 students. 4 are girls. How many are boys?"

I can draw and tell a story. Can you solve?

Mama and I ate a snack. There were 10 things on the table. Then, I dropped my fork on the floor. How many things are still on the table?

Lesson 28: Act out *result unknown* story problems without equations.

113

Name _____   Date _____

Make up a story about 10 things in your house. Draw a picture to go with your story.   Be ready to share your story at school.

Lesson 28:   Act out *result unknown* story problems without equations.

115

Count the dots. Write how many. Draw the same number of dots below but going up and down instead of across.

I can show a number by writing the numeral or by drawing dots.

I can draw 8 circles. I have 8. One more is 9.

Make your own 5-group cards! Cut the cards out on the dotted lines. On one side, write the numbers from 1 to 10. On the other side, show the 5-group dot picture that goes with the number.

This is just like the Math Way of counting on my fingers! I have a row of 5 dots and then 1 more. I show 6 as 5 fingers on one hand and 1 on the other. I can count: five, six.

EUREKA MATH

Lesson 29:   Order and match numeral and dot cards from 1 to 10. State 1 more than a given number.

117

© 2018 Great Minds®. eureka-math.org

Name _____   Date _____

Count the dots. Write how many. Draw the same number of dots below, but going up and down instead of across. The number 6 has been done for you.

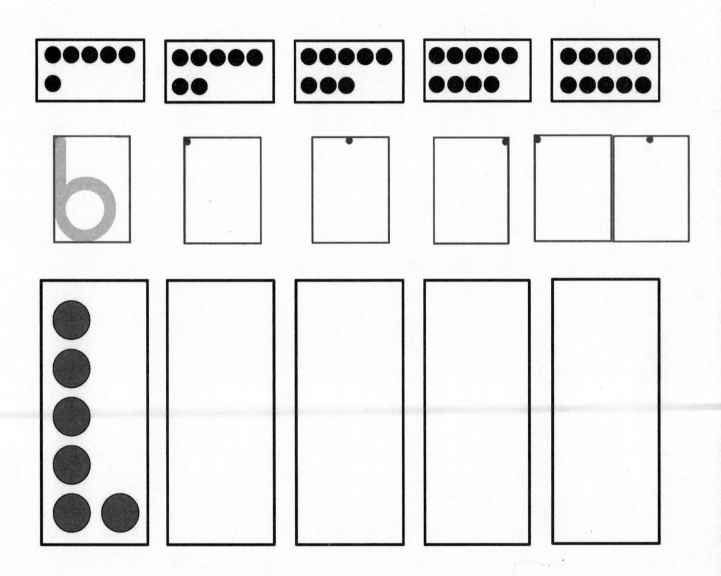

Lesson 29: Order and match numeral and dot cards from 1 to 10. State 1 more than a given number.

© 2018 Great Minds®. eureka-math.org

119

Make your own 5-group cards! Cut the cards out on the dotted lines. On one side, write the numbers from 1 to 10. On the other side, show the 5-group dot picture that goes with the number.

Lesson 29:   Order and match numeral and dot cards from 1 to 10. State 1 more than a given number.

© 2018 Great Minds®. eureka-math.org

EUREKA MATH

Draw the missing stairs. Write the numbers below each step.

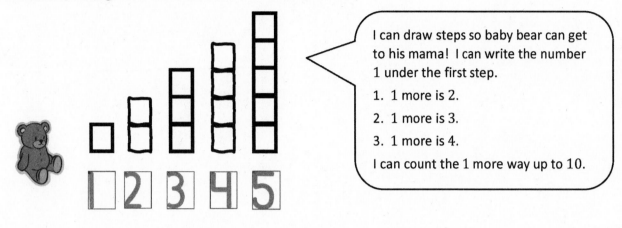

I can draw steps so baby bear can get to his mama! I can write the number 1 under the first step.

1. 1 more is 2.

2. 1 more is 3.

3. 1 more is 4.

I can count the 1 more way up to 10.

Draw 1 more cube on each stair so the cubes match the number. Say as you draw, "1. One more is two. 2. One more is three."

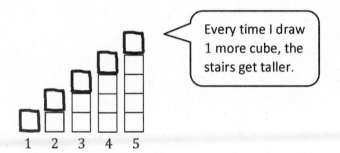

Every time I draw 1 more cube, the stairs get taller.

1   2   3   4   5

 EUREKA MATH®

Lesson 30:   Make math stairs from 1 to 10 in cooperative groups.

121

© 2018 Great Minds®. eureka-math.org

Name _____ Date _____

Draw the missing stairs. Write the numbers below each step.

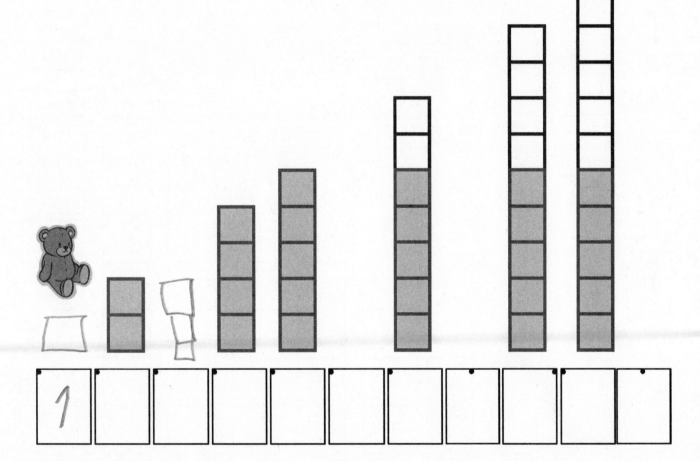

Ask someone to help you write about what you think baby bear will do now that you have helped him to get home. Use the back of this paper.

Lesson 30: Make math stairs from 1 to 10 in cooperative groups.

123

Draw 1 more cube on each stair so the cubes match the number. Say as you draw, "1. One more is two. 2. One more is three."

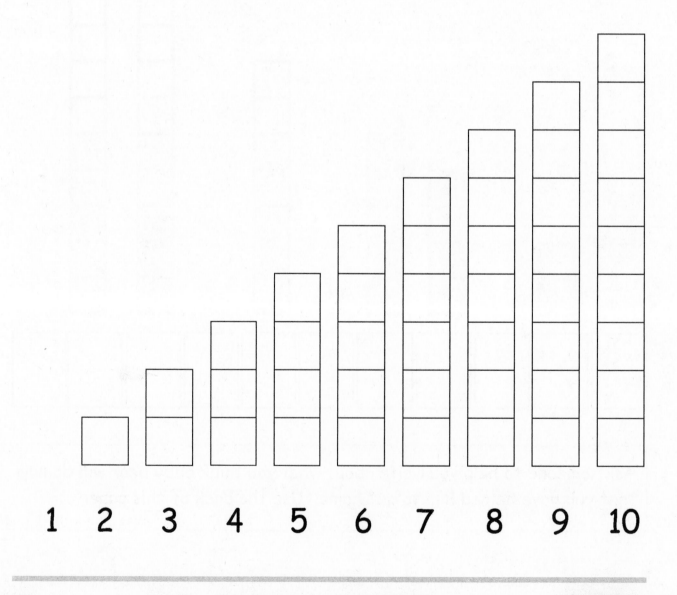

1 2 3 4 5 6 7 8 9 10

Draw one more circle. Color all the circles, and write how many.

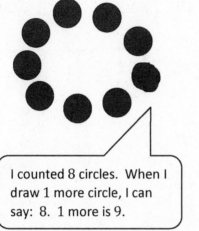

I counted 8 circles. When I draw 1 more circle, I can say: 8. 1 more is 9.

Draw one more star. Color all the stars, and write how many.

I counted 6 stars. Then, I can say: 6. 1 more is 7.

EUREKA MATH®

Lesson 31:   Arrange, analyze, and draw 1 more up to 10 in configurations other than towers.

125

Name _____   Date _____

Draw one more square. Color all the squares, and write how many.

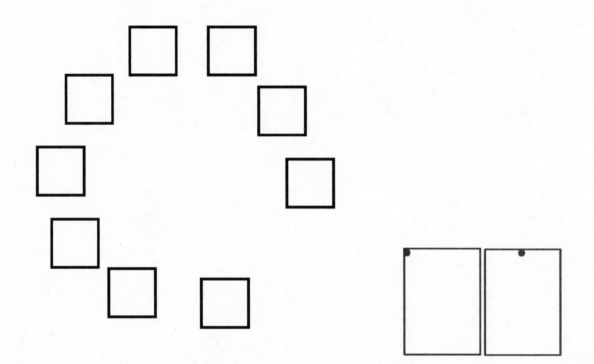

Draw one more cloud. Color all the clouds, and write how many.

Lesson 31:   Arrange, analyze, and draw 1 more up to 10 in configurations other than towers.

127

Write the missing numbers.

2, , 5, 6, 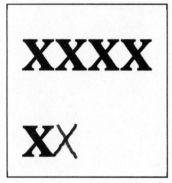, 10

> Each number in the row is 1 more.
> 6. 1 more is 7. Then 8. Then 9.

Draw X's or O's to show 1 more.

> I don't have to start counting at 1 every time. I know there are 3 O's. 1 more is 4. If I drew the O's in a line, there would still be 4 of them.

Tell someone a story about "1 more...and then 1 more." Draw a picture about your story.

> Listen to my story: I have 3 apples in a basket. I put 1 more apple in my basket. 3. 1 more is 4. Then, I put 1 more in my basket. 4. 1 more is 5. I have 5 apples now!

Lesson 32:   Arrange, analyze, and draw sequences of quantities of 1 more, beginning with numbers other than 1.

© 2018 Great Minds®. eureka-math.org

129

Name _____   Date _____

Write the missing numbers.

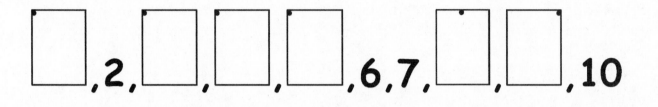

☐ ,2, ☐ ☐ , ☐ ☐ ,6,7, ☐ ☐ ,10

Draw X's or O's to show 1 more.

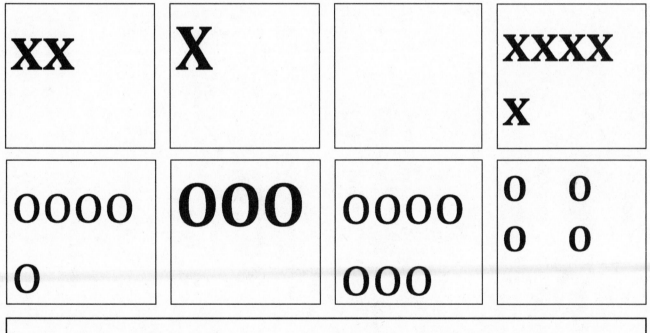

| XX | X | | XXXX X |
| OOOO O | OOO | OOOO OOO | O O O O |

Tell someone a story about "1 more...and then 1 more." Draw a picture about your story.

Lesson 32:   Arrange, analyze, and draw sequences of quantities of 1 more,
             beginning with numbers other than 1.

131

Make 5-Group Cards: Cut the cards out on the dotted lines. On one side, write the numbers from 1-10. On the other side, show the 5-group dot picture that goes with the number. Mix up your cards, and practice putting them in order the "1 less way."

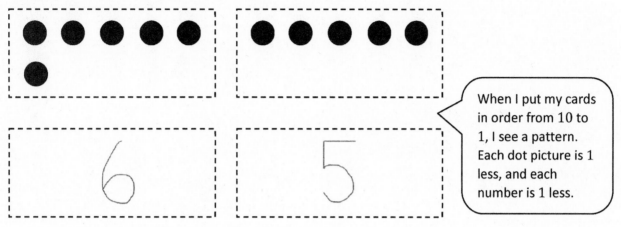

> When I put my cards in order from 10 to 1, I see a pattern. Each dot picture is 1 less, and each number is 1 less.

Make 5-group Cards

Cut the cards out on the dotted lines. On one side, write the numbers from 1-10. On the other side, show the 5-group dot picture that goes with the number. Mix up your cards, and practice putting them in order in the "1 less" way.

Count and color the triangles. Draw a group of triangles that is 1 less. Write how many you drew.

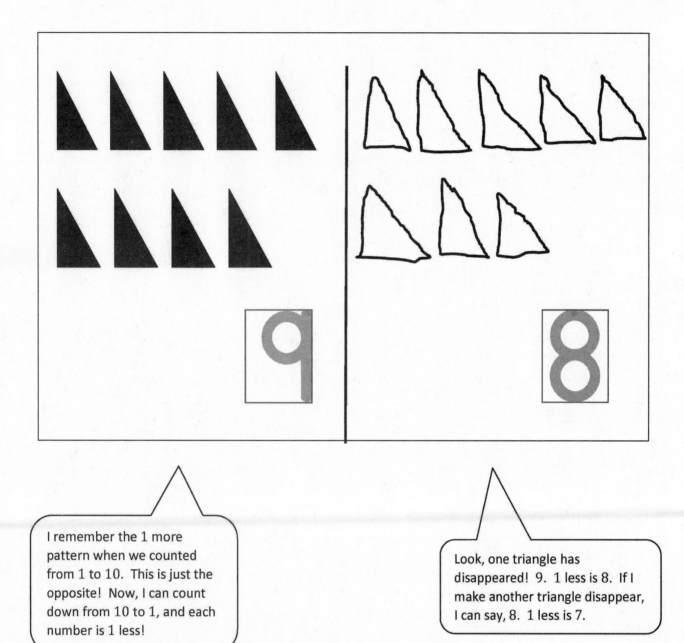

I remember the 1 more pattern when we counted from 1 to 10. This is just the opposite! Now, I can count down from 10 to 1, and each number is 1 less!

Look, one triangle has disappeared! 9. 1 less is 8. If I make another triangle disappear, I can say, 8. 1 less is 7.

Name _____ Date _____

Count and write the number of objects. Draw and write the number of objects that is 1 less.

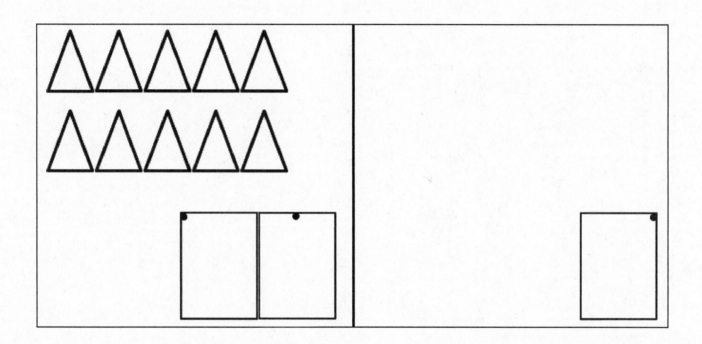

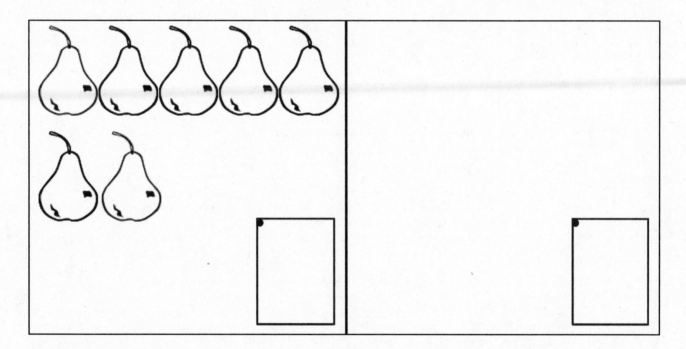

Lesson 34:   Count down from 10 to 1, and state 1 less than a given number.

139

Count all the squares in each tower, and write how many. Share with someone what you notice!

I can count the squares in this tower. There are 4. 1 less is 3.

The towers keep getting smaller and so do the numbers!

I can count the squares in the last tower, too. There is 1 square left. 1 less is 0.

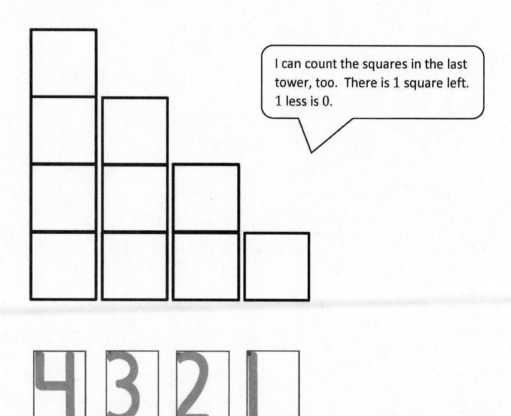

Name _____   Date _____

Count all the squares in each tower, and write how many. Share with
someone what you notice!

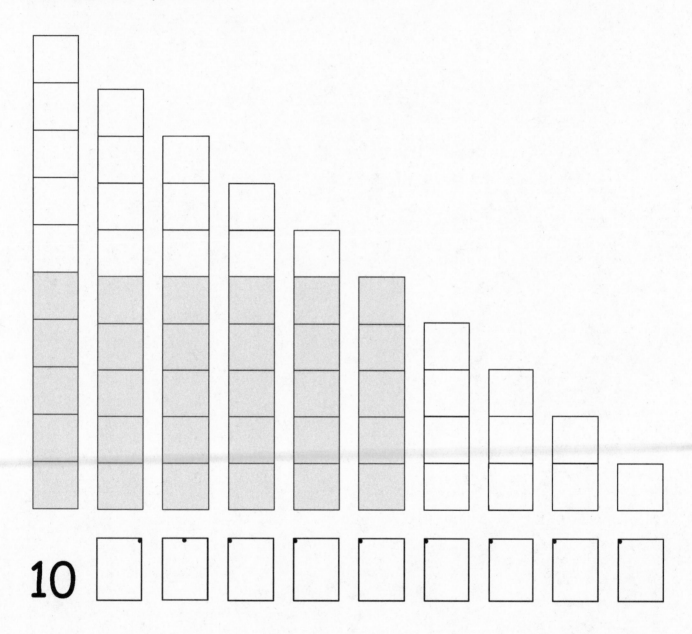

10

EUREKA
MATH

Draw bracelets with the number of beads shown. Write the missing number. Hint: The missing number is 1 less!

I had 8 beads. I know that 1 less is 7. I can call this my 7 bracelet! The next one will be my 6 bracelet. Each bracelet has 1 less.

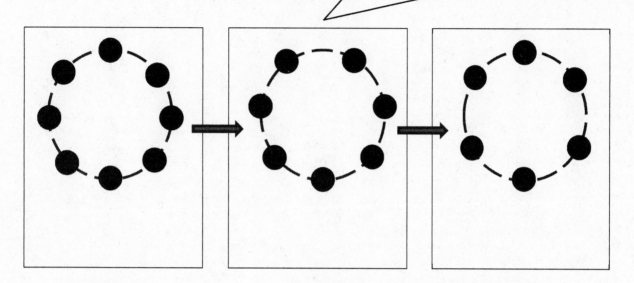

I can count down from 10 to 0. When I start at 10, I know that the next number will be 1 less.

10, 9, 8, 7, 6, 5, 4, 3, 2, 1, 0

Lesson 36: Arrange, analyze, and draw sequences of quantities that are 1 less in configurations other than towers.

145

© 2018 Great Minds®. eureka-math.org

Name _____   Date _____

Draw bracelets with the number of beads shown.

Write the missing number. Hint: The missing number is 1 less!

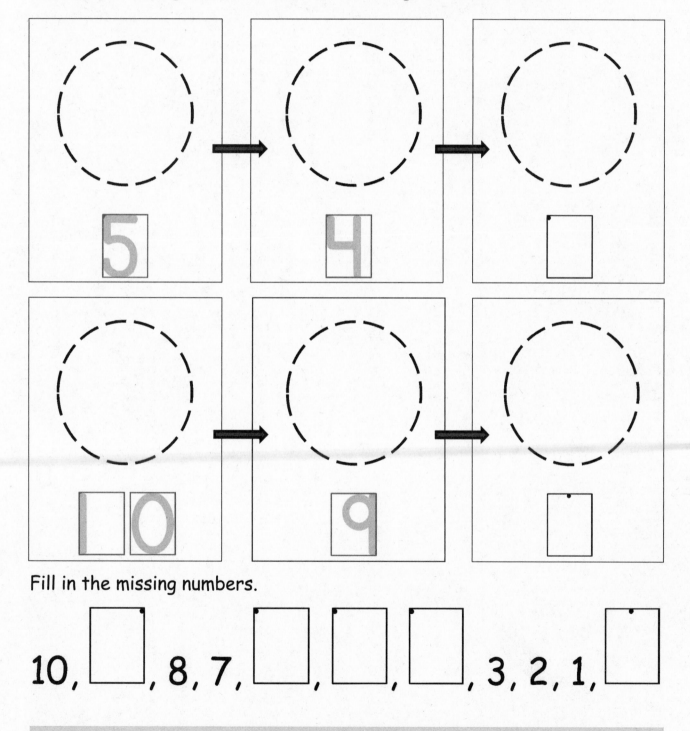

Fill in the missing numbers.

10, ☐, 8, 7, ☐, ☐, ☐, 3, 2, 1, ☐

Lesson 36:   Arrange, analyze, and draw sequences of quantities that are 1 less in
configurations other than towers.

147

© 2018 Great Minds®. eureka-math.org

Note: Be sure to ask your child about his/her mystery number from today's Number Fair!

Count how many are in each group. Write the number in the box. Circle the smaller group.

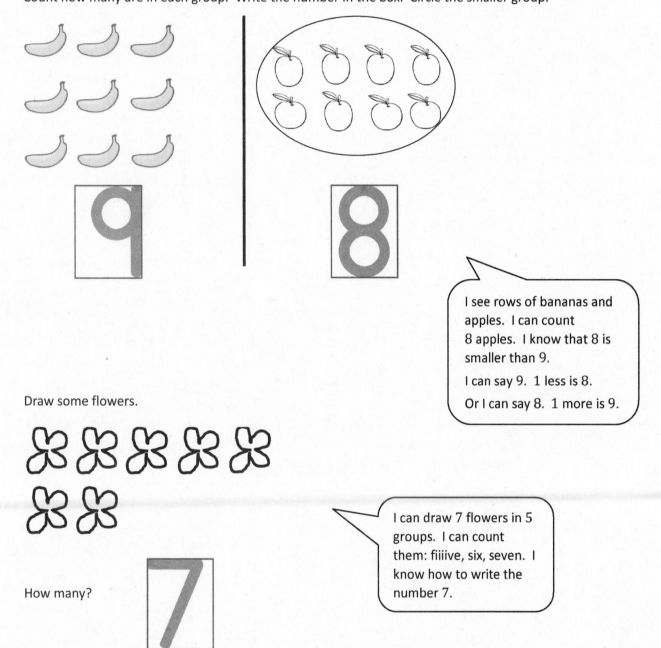

I see rows of bananas and apples. I can count 8 apples. I know that 8 is smaller than 9.

I can say 9. 1 less is 8.

Or I can say 8. 1 more is 9.

Draw some flowers.

How many?

I can draw 7 flowers in 5 groups. I can count them: fiiiive, six, seven. I know how to write the number 7.

Name _____   Date _____

Count how many are in each group. Write the number.

CHALLENGE: Circle the smaller group in each row.

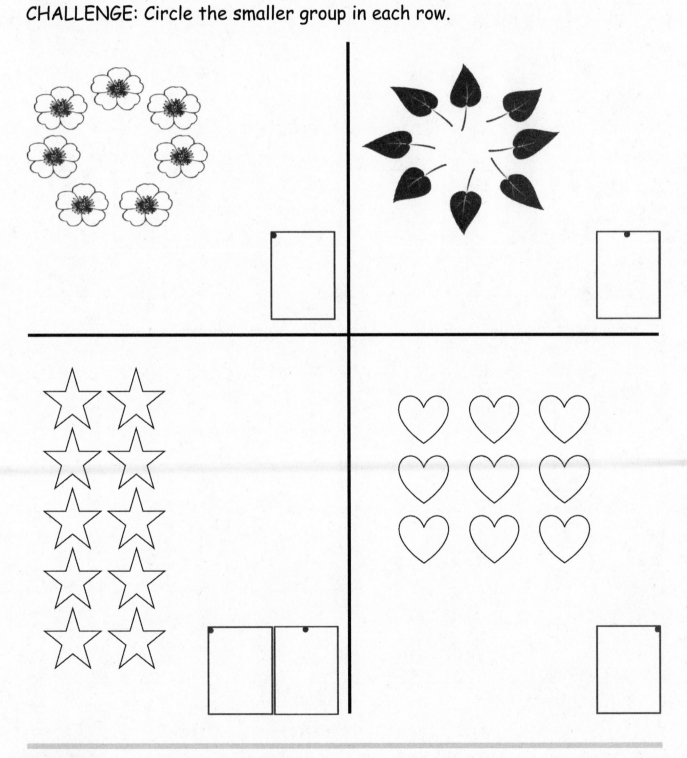

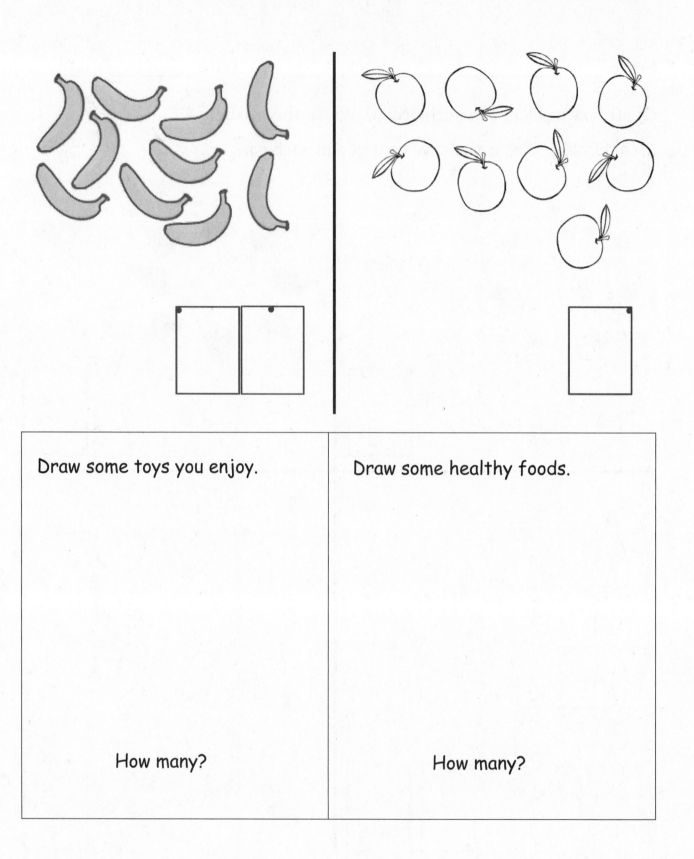

Draw some toys you enjoy.

How many?

Draw some healthy foods.

How many?

Grade K
Module 2

Draw a line from the shape to its matching object.

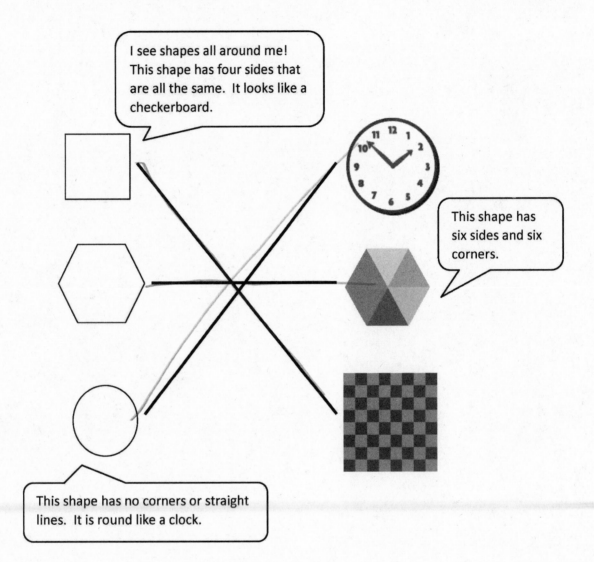

I see shapes all around me! This shape has four sides that are all the same. It looks like a checkerboard.

This shape has six sides and six corners.

This shape has no corners or straight lines. It is round like a clock.

Lesson 1: Find and describe flat triangles, squares, rectangles, hexagons, and circles using informal language without naming.

155

© 2018 Great Minds®. eureka-math.org

Name _____   Date _____

Draw a line from the shape to its matching object.

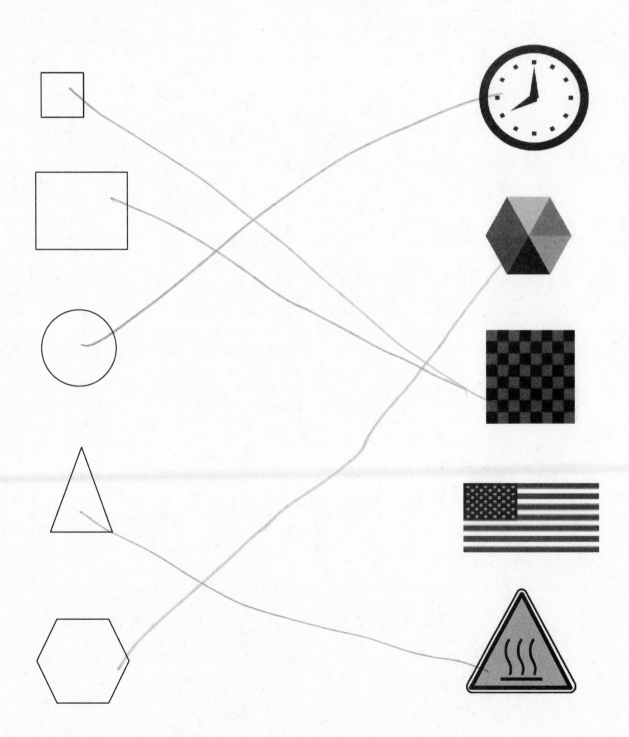

Lesson 1:   Find and describe flat triangles, squares, rectangles, hexagons, and
circles using informal language without naming.

157

© 2018 Great Minds®. eureka-math.org

Color the triangles red and the other shapes blue.

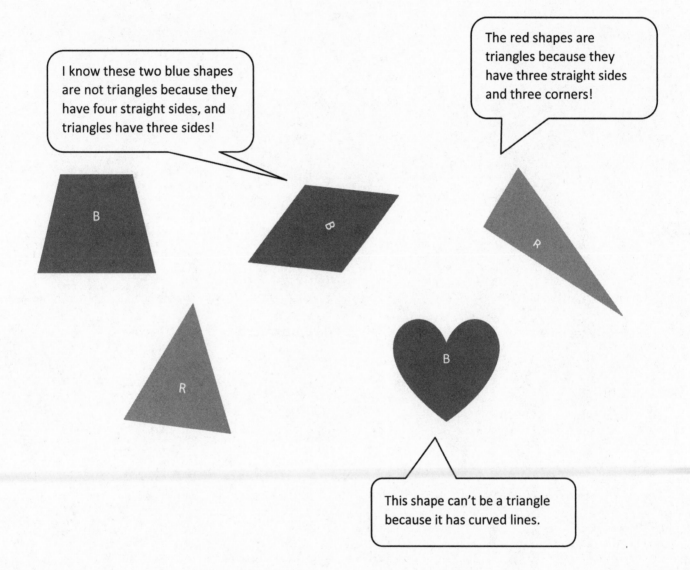

Lesson 2: Explain decisions about classifications of triangles into categories using variants and non examples. Identify shapes as triangles.

159

Name _____   Date _____

Color the triangles red and the other shapes blue.

Draw 2 different triangles of your own.

Lesson 2:   Explain decisions about classifications of triangles into categories using variants and non examples. Identify shapes as triangles.

© 2018 Great Minds®. eureka-math.org

161

Color all the rectangles red. Color all the triangles green.

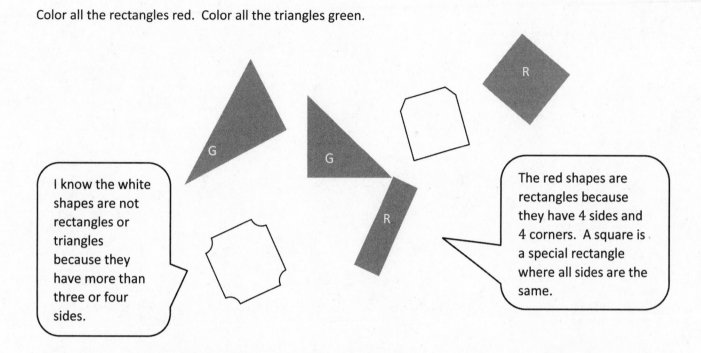

I know the white shapes are not rectangles or triangles because they have more than three or four sides.

The red shapes are rectangles because they have 4 sides and 4 corners. A square is a special rectangle where all sides are the same.

In the box, draw 2 rectangles and 2 triangles. How many shapes did you draw? Put your answer in the circle.

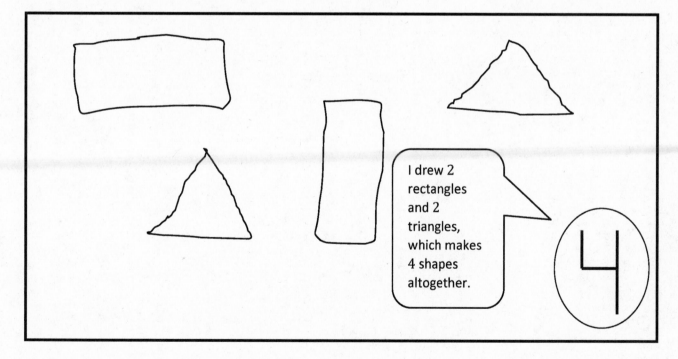

I drew 2 rectangles and 2 triangles, which makes 4 shapes altogether.

Lesson 3:   Explain decisions about classifications of rectangles into categories using variants and non examples. Identify shapes as rectangles.

163

© 2018 Great Minds®. eureka-math.org

Name _____ Date _____

Color all the rectangles red. Color all the triangles green.

On the back of your paper, draw 2 rectangles and 3 triangles.

How many shapes did you draw? Put your answer in the circle.

EUREKA
MATH®

Lesson 3:   Explain decisions about classifications of rectangles into categories
             using variants and non examples. Identify shapes as rectangles.

© 2018 Great Minds®. eureka-math.org

165

Color the triangles blue.

Color the rectangles red.

Color the circles green.

Color the hexagons yellow.

This looks similar to a circle because it has curved lines and no corners. I know it is not a circle because it looks like it is stretched out.

R

B

R

Y

G

Y

Y

I know this isn't a shape because it is not closed.

This shape does not look like a typical hexagon, but it has six straight sides, so it is a hexagon!

Lesson 4:   Explain decisions about classifications of hexagons and circles, and identify them by name. Make observations using variants and non-examples.

167

© 2018 Great Minds®. eureka-math.org

Name _____   Date _____

Color the triangles blue.

Color the rectangles red.

Color the circles green.

Color the hexagons yellow.

On the back of your paper, draw 2 triangles and 1 hexagon.

How many shapes did you draw? _____

     Lesson 4:   Explain decisions about classifications of hexagons ℓend circles, and identify   169
them by name. Make observations using variants and non-examples.

© 2018 Great Minds®. eureka-math.org

Next to the flower, draw a shape with 4 sides, 2 long and 2 short. Color it green.

Below the flower, draw a shape with no corners. Color it red.

Above the flower, draw a shape with 3 straight sides. Color it blue.

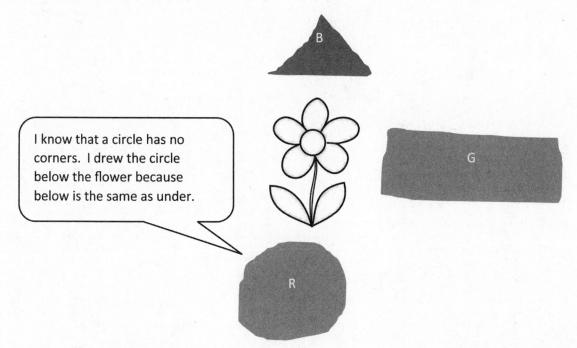

I know that a circle has no corners. I drew the circle below the flower because below is the same as under.

In the box, draw 3 circles and 2 triangles. How many shapes did you draw? Put your answer in the circle.

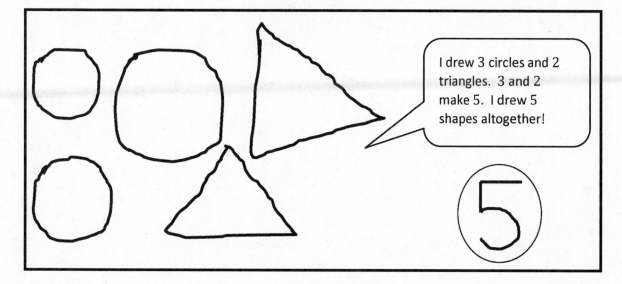

I drew 3 circles and 2 triangles. 3 and 2 make 5. I drew 5 shapes altogether!

EUREKA MATH®

Lesson 5: Describe and communicate positions of all flat shapes using words
above, below, beside, in front of, next to, and *behind.*

171

© 2018 Great Minds®. eureka-math.org

Name _____   Date _____

- **Behind** the elephant, draw a shape with 4 straight sides that are exactly the same length. Color it blue.

- **Above** the elephant, draw a shape with no corners. Color it yellow.

- **In front of** the elephant, draw a shape with 3 straight sides. Color it green.

- **Below** the elephant, draw a shape with 4 sides, 2 long and 2 short. Color it red.

- **Below** the elephant, draw a shape with 6 corners. Color it orange.

On the back of your paper, draw 1 hexagon and 4 triangles.
How many shapes did you draw? Put your answer in the circle.

Lesson 5:   Describe and communicate positions of all flat shapes using the words
              above, below, beside, in front of, next to, and *behind.*   173

© 2018 Great Minds®. eureka-math.org

Find things in your house or in a magazine that look like these solids. Draw the solids or cut out and paste pictures from a magazine.

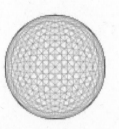

I know this shape! It is pointy at the end and holds ice cream!

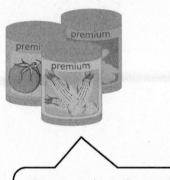

These cans look the same as the solid shape because they both curve in the middle and have circles on the ends.

Name _____   Date _____

Find things in your house or in a magazine that look like these solids. Draw the solids or cut out and paste pictures from a magazine.

Lesson 6:   Find and describe solid shapes using informal language without naming.

177

Cut one set of solid shapes. Sort the 4 solid shapes. Paste them onto the chart.

These shapes roll.   These shapes do not roll.

The cube does not have any curved sides. I know that it will not roll no matter which side I put it down on.

These shapes have circle faces.   These shapes do not have circle faces.

The cylinder has 2 circle faces, and the cone has 1 circle face.

Lesson 7:   Explain decisions about classification of solid shapes into categories.
Name the solid shapes.

179

© 2018 Great Minds®. eureka-math.org

Name _____   Date _____

Cut one set of solid shapes. Sort the 4 solid shapes. Paste onto the chart.

These have corners.   These do not have corners.

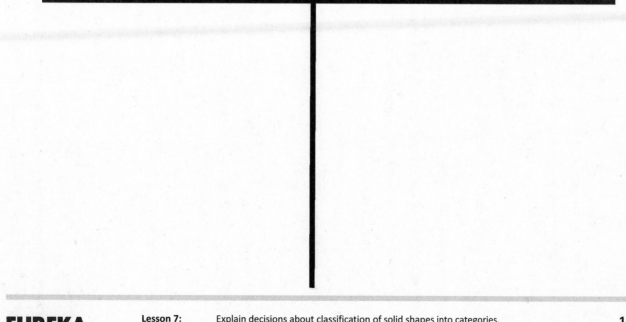

Cut the other set of solid shapes, and make a rule for how you sorted
them. Paste onto the chart.

EUREKA
MATH®

Lesson 7:   Explain decisions about classification of solid shapes into categories.
              Name the solid shapes.

© 2018 Great Minds®. eureka-math.org

181

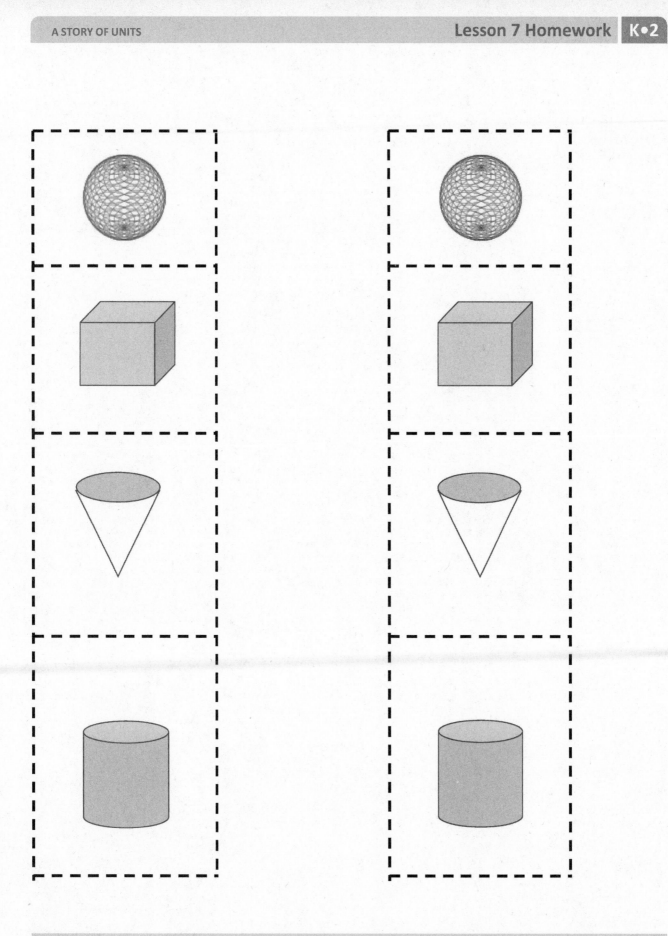

Lesson 7:   Explain decisions about classification of solid shapes into categories.
Name the solid shapes.

183

EUREKA
MATH

Tell someone at home the names of each solid shape.

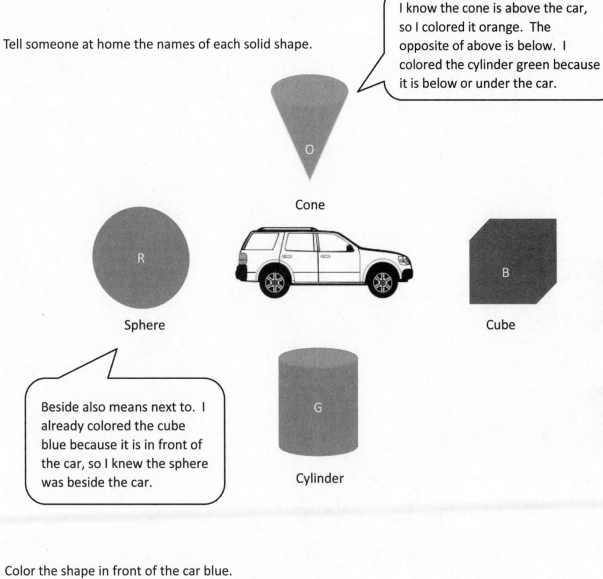

I know the cone is above the car, so I colored it orange. The opposite of above is below. I colored the cylinder green because it is below or under the car.

Cone

Sphere

Cube

Beside also means next to. I already colored the cube blue because it is in front of the car, so I knew the sphere was beside the car.

Cylinder

Color the shape in front of the car blue.

Color the shape above the car orange.

Color the shape below the car green.

Color the shape beside the car red.

Lesson 8: Describe and communicate positions of all solid shapes using the words *above, below, beside, in front of, next to,* and *behind.*

185

Name _____ Date _____

Tell someone at home the names of each solid shape.

Sphere   Cylinder   Cone   Cube

Color the car **beside** the stop sign green.

Circle the **next** car with blue.

Color the car **behind** the circled car red.

Draw a road **below** the cars.

Draw a policeman **in front of** the cars.

Draw a sun **above** the cars.

 EUREKA MATH®

Lesson 8:   Describe and communicate positions of all solid shapes using the
words *above, below, beside, in front of, next to,* and *behind.*

187

© 2018 Great Minds®. eureka-math.org

In each row, circle the one that doesn't belong. Explain your choice to a grown-up.

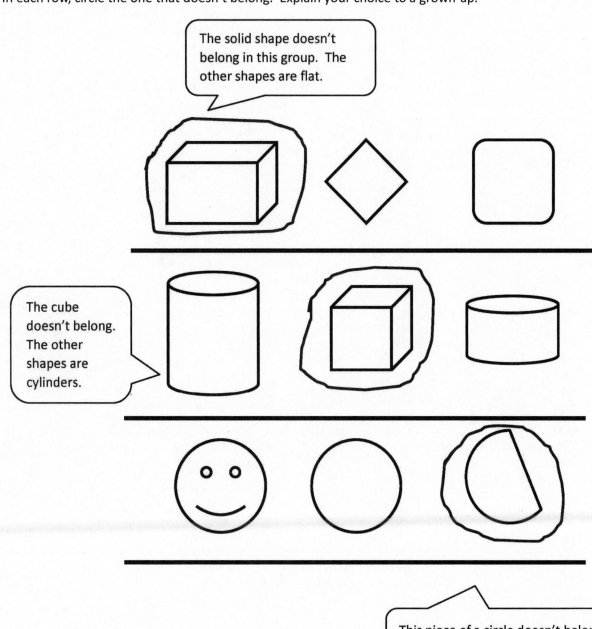

The solid shape doesn't belong in this group. The other shapes are flat.

The cube doesn't belong. The other shapes are cylinders.

This piece of a circle doesn't belong. The other shapes really are circles.

Lesson 9:   Identify and sort shapes as two-dimensional or three-dimensional, and recognize two-dimensional and three-dimensional shapes in different orientations and sizes.

© 2018 Great Minds®. eureka-math.org

189

Name _____   Date _____

In each row, circle the one that doesn't belong. Explain your choice to a grown-up.

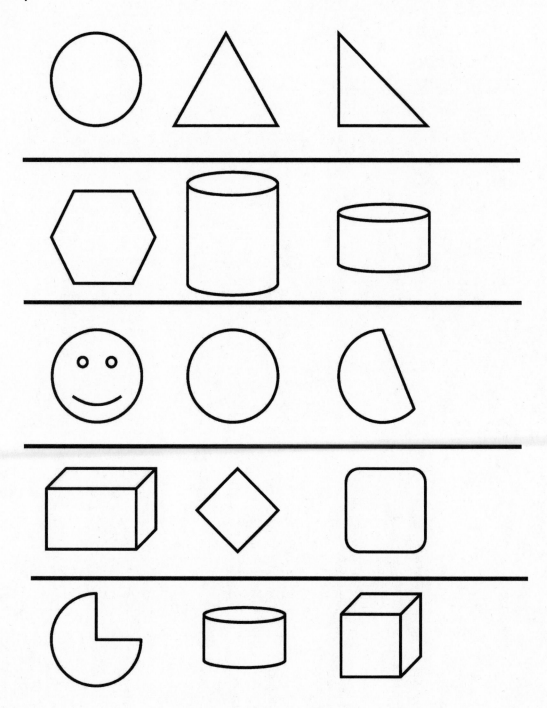

Search your house to see what shapes and solids you can find. Draw the shapes that you see by tracing the faces of the solids that you find. Color your collage.

This is a block from my room! It is a cube, which has 6 square faces.

I traced a can of green beans. The shape of the can is a cylinder, and the face is a circle.

I ate an ice cream cone for dessert.

Its face is a circle, too!

This is my bouncy ball! It is a sphere. It doesn't have a face. It's curved all over.

Lesson 10:   Culminating task—collaborative groups create displays of different flat shapes with examples, non-examples, and a corresponding solid shape.

© 2018 Great Minds®. eureka-math.org

193

Name _____   Date _____

Shape Up Your Kitchen!

Search your kitchen to see what shapes and solids you can find. Make a kitchen-shaped collage by drawing the shapes that you see and by tracing the faces of the solids that you find. Color your collage.

Grade K
Module 3

Draw 2 more trees that are shorter than these trees.

Count how many trees you have now.

Write the number in the box.

On the back of your paper, draw something that is shorter than the refrigerator.

My kitty stands beside the refrigerator. The refrigerator is so tall! Kitty is much shorter than the refrigerator.

Lesson 1:   Compare lengths using *taller than* and *shorter than* with aligned and non-aligned endpoints.

© 2018 Great Minds®. eureka-math.org

199

Name _____ Date _____

Draw 3 more flowers that are shorter than these flowers.

Count how many flowers you have now. Write the number in the box.

Draw 2 more ladybugs that are taller than these ladybugs.

Count how many ladybugs you have now. Write the number in the box.

On the back of your paper, draw something that is taller than you. Draw something that is shorter than a flagpole.

EUREKA MATH

Lesson 1:   Compare lengths using *taller than* and *shorter than* with aligned and non-aligned endpoints.

201

© 2018 Great Minds®. eureka-math.org

Using the 1-foot piece of string from class, find three items at home that are shorter than your piece of string and three items that are longer than your piece of string. Draw a picture of those objects on the chart. Try to find at least one thing that is about the same length as your string, and draw a picture of it on the back.

Shorter than the string

Longer than the string

My building block, my truck, and Sam's sippy cup are all shorter than the string.

I compare the string to things in my room. My bed, my rug, and my jump rope are longer than the string!

Name _____ Date _____

Using the piece of string from class, find three items at home that are shorter than your piece of string and three items that are longer than your piece of string. Draw a picture of those objects on the chart. Try to find at least one thing that is about the same length as your string, and draw a picture of it on the back.

Shorter than the string	Longer than the string

Take out a new crayon. Use a red crayon to circle objects with lengths shorter than the crayon. Use a blue crayon to circle objects with lengths longer than the crayon.

I can use *longer than* and *shorter than* as I compare lengths.

I compare the length of my crayon with the length of this shape. My crayon is shorter.

Lesson 3:   Make a series of *longer than* and *shorter than* comparisons.

207

© 2018 Great Minds®. eureka-math.org

Name _____ Date _____

Take out a new crayon. Circle objects with lengths shorter than the crayon blue. Circle objects with lengths longer than the crayon red.

On the back of your paper, draw some things shorter than the crayon and longer than the crayon. Draw something that is the same length as the crayon.

Lesson 3:   Make a series of *longer than* and *shorter than* comparisons.

209

© 2018 Great Minds®. eureka-math.org

Use a red crayon to circle the sticks that are shorter than the 5-stick.

I count 5 cubes on this 5-stick.

This stick has 3 cubes. It is shorter than the 5-stick.

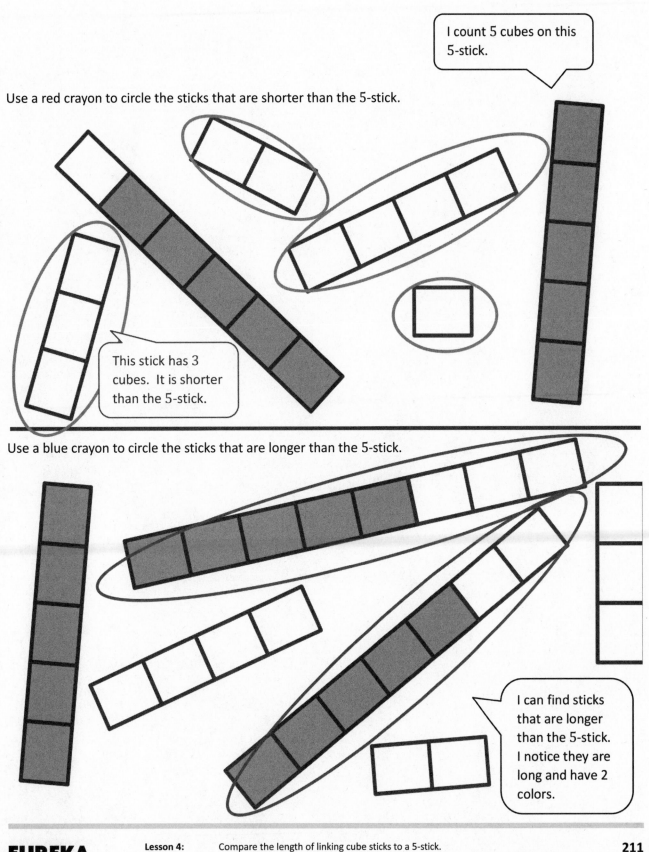

Use a blue crayon to circle the sticks that are longer than the 5-stick.

I can find sticks that are longer than the 5-stick. I notice they are long and have 2 colors.

Lesson 4:   Compare the length of linking cube sticks to a 5-stick.

EUREKA MATH

© 2018 Great Minds®. eureka-math.org

211

Name _____   Date _____

Use a red crayon to circle the sticks that are shorter than the 5-stick.

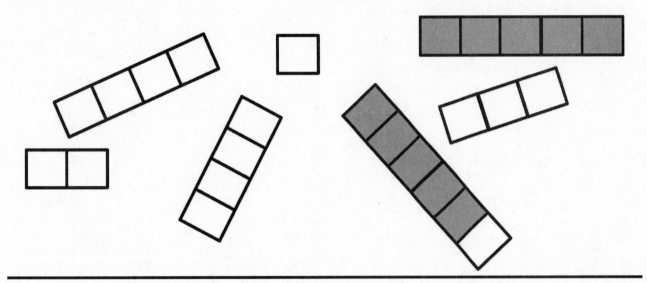

Use a blue crayon to circle the sticks that are longer than the 5-stick.

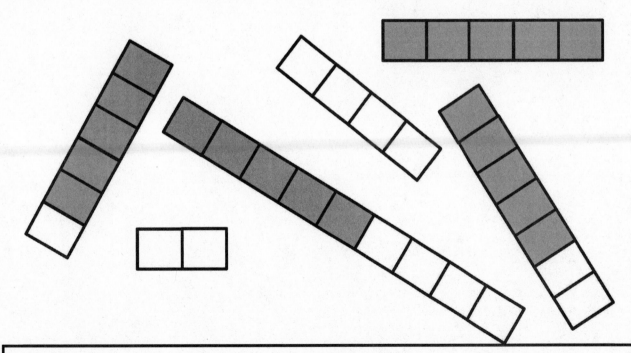

On the back, draw a 7-stick. Draw a stick longer than it. Draw a stick shorter than it.

Lesson 4:   Compare the length of linking cube sticks to a 5-stick.

213

© 2018 Great Minds®. eureka-math.org

Circle the stick that is shorter than the other.

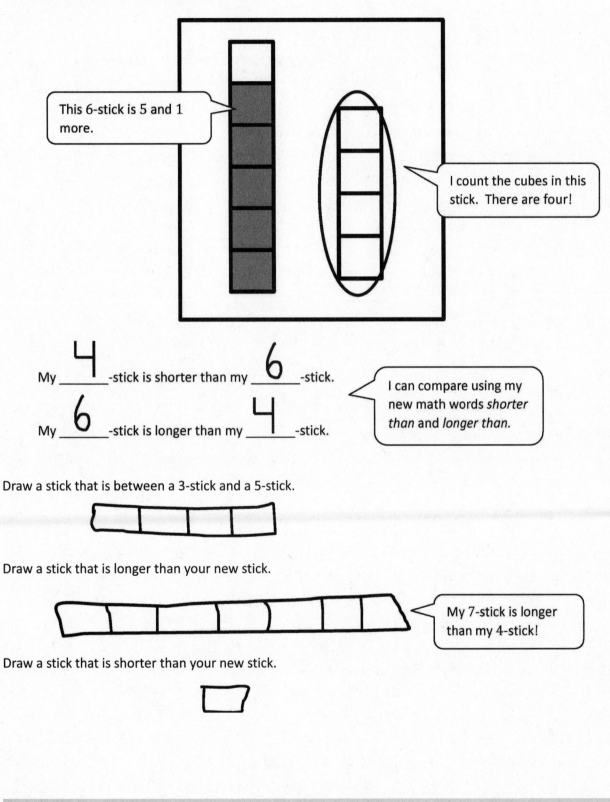

This 6-stick is 5 and 1 more.

I count the cubes in this stick. There are four!

My ___4___ -stick is shorter than my ___6___ -stick.

My ___6___ -stick is longer than my ___4___ -stick.

I can compare using my new math words *shorter than* and *longer than*.

Draw a stick that is between a 3-stick and a 5-stick.

Draw a stick that is longer than your new stick.

My 7-stick is longer than my 4-stick!

Draw a stick that is shorter than your new stick.

EUREKA MATH

Lesson 5: Determine which linking cube stick is *longer than* or *shorter than* the other. 215

© 2018 Great Minds®. eureka-math.org

Name _____   Date _____

Circle the stick that is shorter than the other.

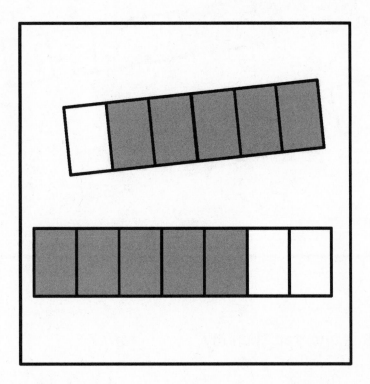

My _____ -stick is shorter than my _____ -stick.

My _____ -stick is longer than my _____ -stick.

On the back of your paper, draw a 7-stick.

Draw a stick that is longer than the 7-stick.

Draw a stick that is shorter than the 7-stick.

EUREKA
MATH®

Circle the stick that is longer than the other.

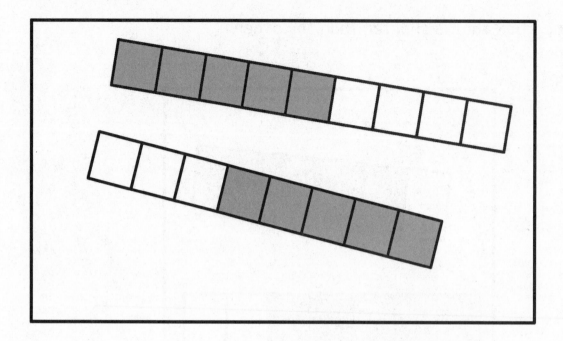

My _____ -stick is shorter than my _____ -stick.

My _____ -stick is longer than my _____ -stick.

On the back of your paper, draw a stick that is between a 4- and a 6-stick.

Draw a stick that is longer than your new stick.

Draw a stick this is shorter than your new stick.

Lesson 5:   Determine which linking cube stick is *longer than* or *shorter than* the other.

Color the cubes to show the length of the object.

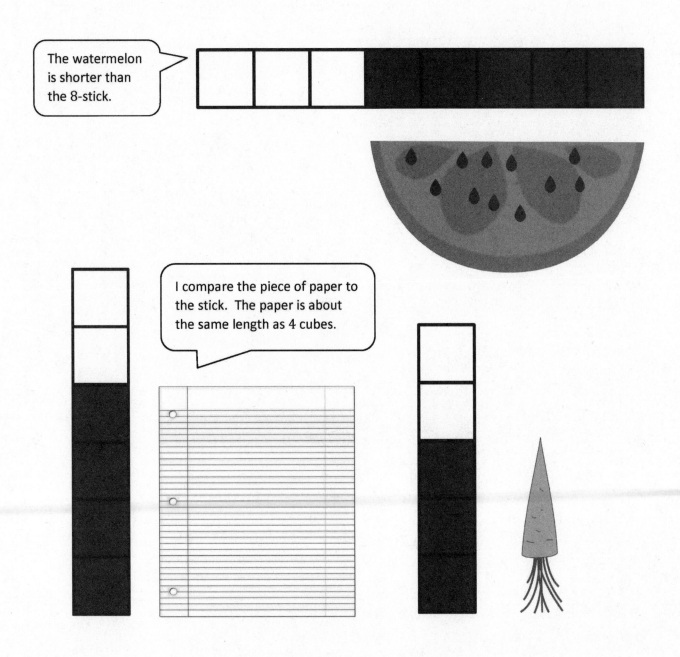

The watermelon is shorter than the 8-stick.

I compare the piece of paper to the stick. The paper is about the same length as 4 cubes.

Name _____   Date _____

Color the cubes to show the length of the object.

Lesson 6:   Compare the length of linking cube sticks to various objects.

© 2018 Great Minds®. eureka-math.org

221

These boxes represent cubes.

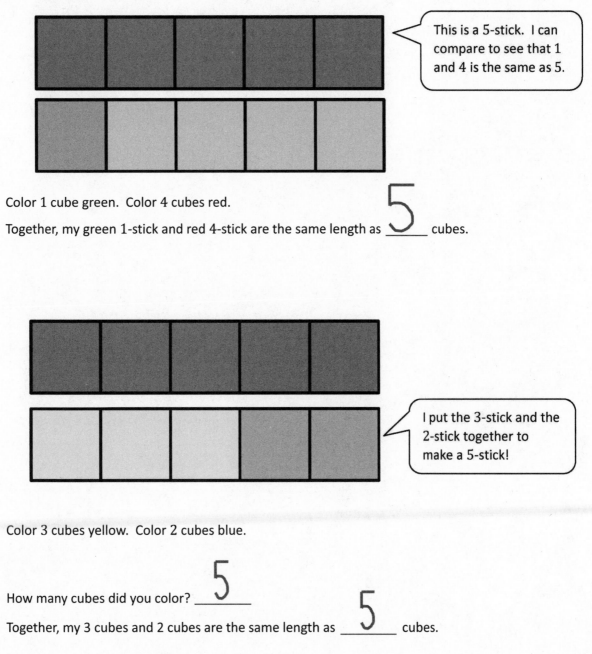

This is a 5-stick. I can compare to see that 1 and 4 is the same as 5.

Color 1 cube green. Color 4 cubes red.

Together, my green 1-stick and red 4-stick are the same length as __5__ cubes.

I put the 3-stick and the 2-stick together to make a 5-stick!

Color 3 cubes yellow. Color 2 cubes blue.

How many cubes did you color? __5__

Together, my 3 cubes and 2 cubes are the same length as __5__ cubes.

Name _____   Date _____

These boxes represent cubes.

Color 2 cubes green. Color 3 cubes blue.

Together, my green 2-stick and blue 3-stick are the same length as 5 cubes.

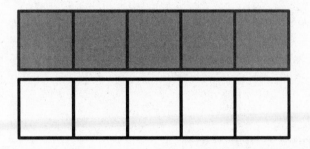

Color 3 cubes blue. Color 2 cubes green.

Together, my blue 3-stick and green 2-stick are the same length

as ___ cubes.

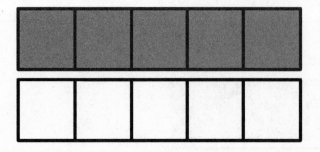

Color 1 cube green. Color 4 cubes blue.

How many did you color? _____

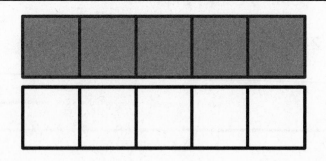

Color 4 cubes green. Color 1 cube blue.

How many did you color? _____

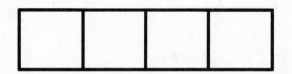

Color 2 cubes yellow. Color 2 cubes blue.

Together, my 2 yellow and 2 blue are the same as _____.

EUREKA MATH®

Draw an object that would be heavier than the one in the picture.

I'd rather carry a pencil in my backpack than a heavy dictionary. The pencil is much lighter!

It's easy for me to pick up the notebook. But, I can't pick up Daddy. He's too heavy.

EUREKA MATH

Lesson 8: Compare using *heavier than* and *lighter than* with classroom objects.

227

© 2018 Great Minds®. eureka-math.org

Name _____   Date _____

Draw an object that would be lighter than the one in the picture.

Draw something inside the box that is heavier than the object on the balance.

> A bowling ball is heavier than a shoe. It takes so many muscles to pick up a heavy bowling ball.

Draw something lighter than the object on the balance.

> An orange is lighter than a pineapple. It's easier to carry. It weighs so little, I can pick up more than one. I can even throw it!

Lesson 9:   Compare objects using *heavier than*, *lighter than*, and *the same as* with balance scales.

231

Name _____   Date _____

Draw something inside the box that is heavier than the object on the balance.

Lesson 9:   Compare objects using *heavier than, lighter than,* and *the same as* with balance scales.

© 2018 Great Minds®. eureka-math.org

233

Draw something lighter than the object on the balance.

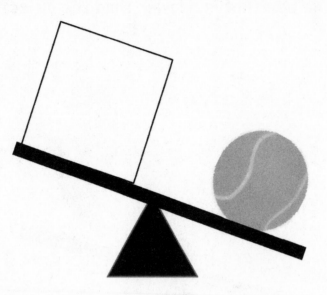

Lesson 9: Compare objects using *heavier than, lighter than,* and *the same as* with balance scales.

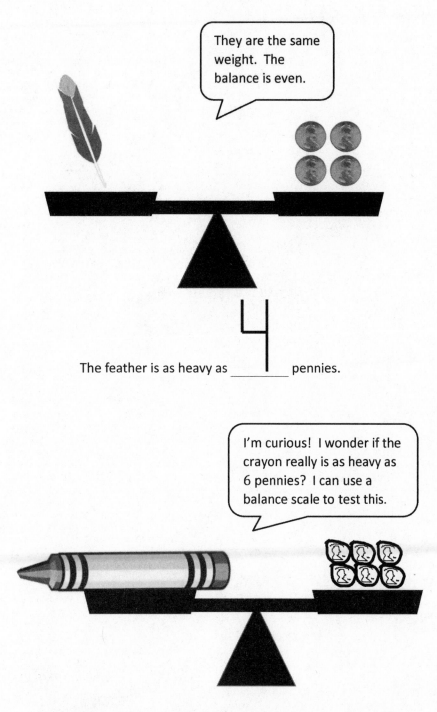

The feather is as heavy as _____ pennies.

Draw in the pennies so that the crayon is as heavy as 6 pennies.

EUREKA MATH

Lesson 10:   Compare the weight of an object to a set of unit weights on a balance scale.

© 2018 Great Minds®. eureka-math.org

235

Name _____   Date _____

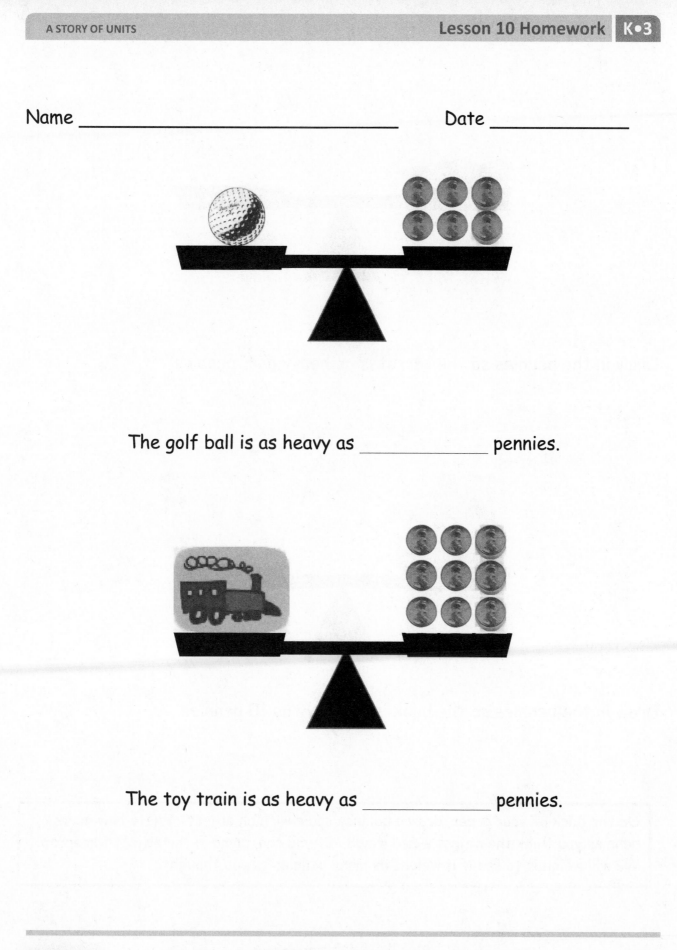

The golf ball is as heavy as _____ pennies.

The toy train is as heavy as _____ pennies.

Lesson 10:   Compare the weight of an object to a set of unit weights on a balance
scale.

© 2018 Great Minds®. eureka-math.org

237

Draw in the pennies so the carrot is as heavy as 5 pennies.

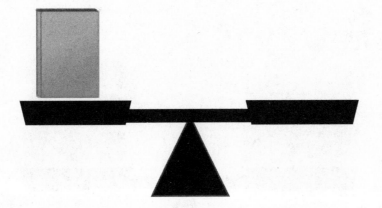

Draw in the pennies so the book is as heavy as 10 pennies.

On the back of your paper, draw a balance scale with an object. Write how many pennies you think the object would weigh. If you can, bring in the object tomorrow. We will weigh it to see if it weighs as many pennies as you thought.

Lesson 10:   Compare the weight of an object to a set of unit weights on a balance scale.

EUREKA MATH®

Draw linking cubes so each side weighs the same.

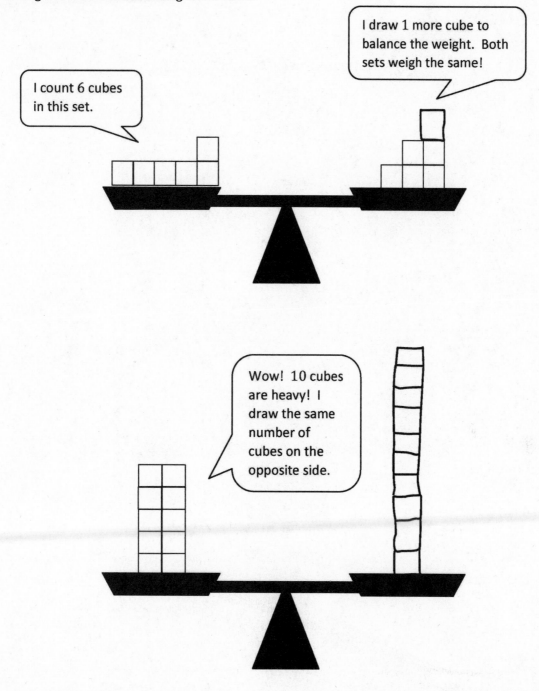

EUREKA
MATH®

Name _____   Date _____

Draw linking cubes so each side weighs the same.

© 2018 Great Minds®. eureka-math.org

The apple is as heavy as _____3_____ cupcakes.

The apple is as heavy as _____2_____ sneakers.

Lesson 12:   Compare the weight of an object with sets of different objects on a
               balance scale.

© 2018 Great Minds®. eureka-math.org

243

Name _____   Date _____

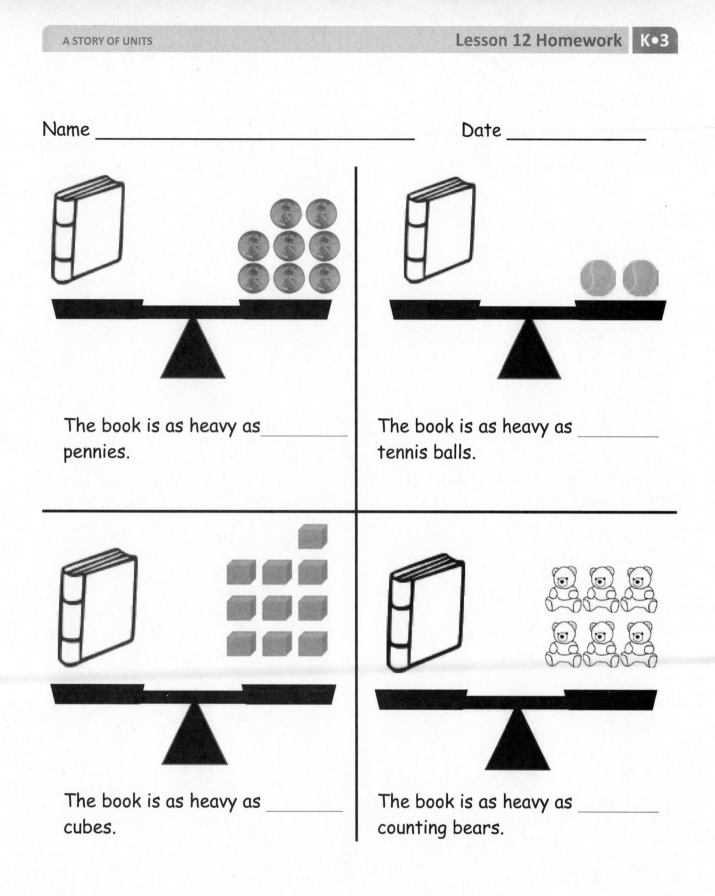

The book is as heavy as_____ pennies.

The book is as heavy as _____ tennis balls.

The book is as heavy as _____ cubes.

The book is as heavy as _____ counting bears.

Lesson 12:   Compare the weight of an object with sets of different objects on a balance scale.

© 2018 Great Minds®. eureka-math.org

245

Each rectangle shows 8 items. Circle two different sets within each. The two sets represent the two parts that make up the 8 objects.

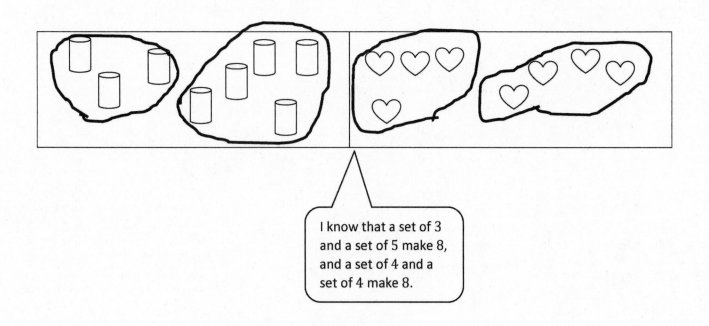

I know that a set of 3 and a set of 5 make 8, and a set of 4 and a set of 4 make 8.

Lesson 13: Compare volume using *more than, less than* and *the same as* by pouring.

247

© 2018 Great Minds®. eureka-math.org

Name _____   Date _____

In class, we have been working on capacity. Encourage your child to explore with different-sized containers to see which ones have the biggest and smallest capacity. Children can experiment by pouring liquid from one container to another.

All the homework you will see for the next few days will be a review of skills taught from Module 1.

Each rectangle shows 6 objects. Circle 2 different sets within each. The two sets represent the two parts that make up the 6 objects. The first one has been done for you.

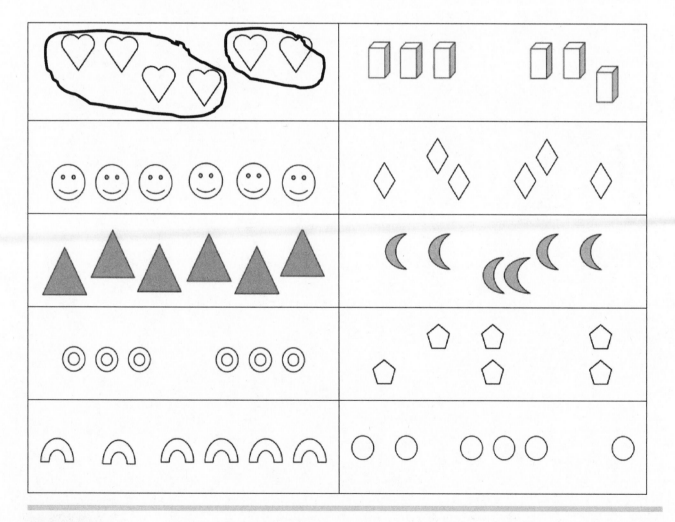

EUREKA
MATH®

© 2018 Great Minds®. eureka-math.org

Within each rectangle, make one set of 8 objects.

There are 9 clouds. I circle 8 of them. It's like 8 is hiding inside of 9.

Here I count 10, and I circle 8 of them.

Name _____   Date _____

Within each rectangle, make one set of 6 objects. The first one has been done for you.

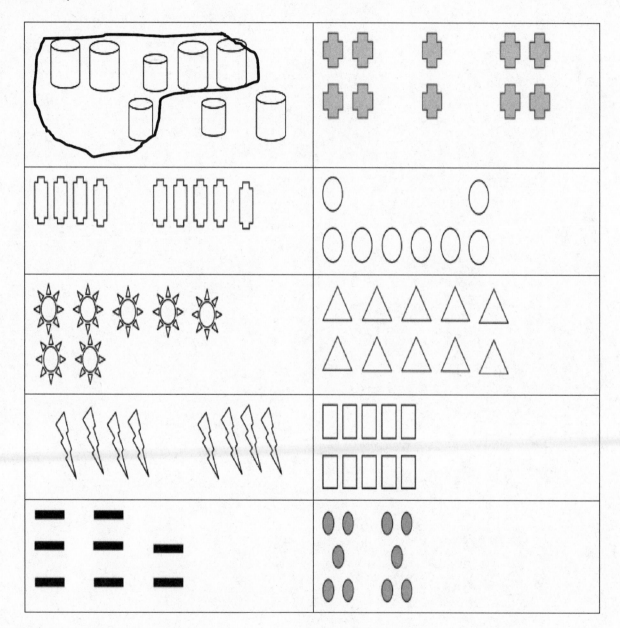

Circle 2 sets within each set of 8.

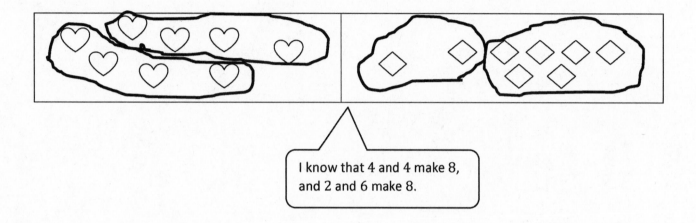

I know that 4 and 4 make 8, and 2 and 6 make 8.

Name _____ Date _____

Circle 2 sets within each set of 7. The first one has been done for you.

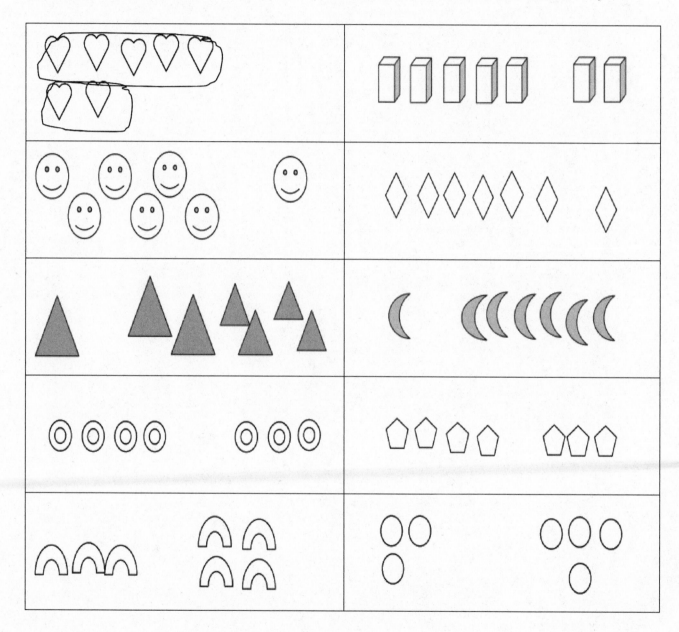

Cover the shape with beans. Count how many, and write the number in the box.

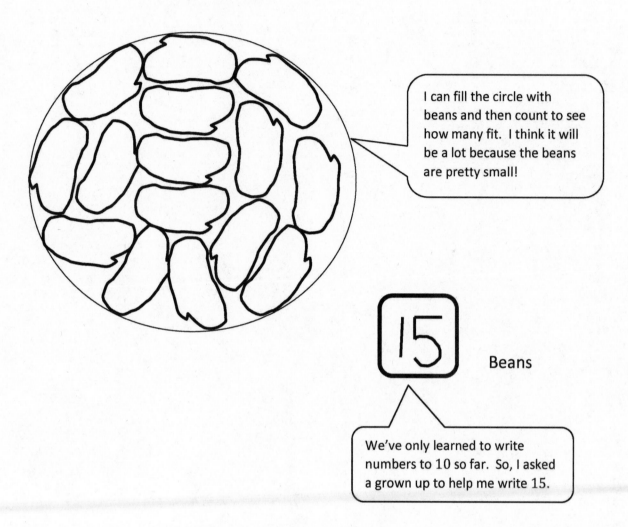

I can fill the circle with beans and then count to see how many fit. I think it will be a lot because the beans are pretty small!

15 Beans

We've only learned to write numbers to 10 so far. So, I asked a grown up to help me write 15.

© 2018 Great Minds®. eureka-math.org

Name _____   Date _____

Trace your hand. Cover the tracing with pennies. Have an adult trace his or her hand. Cover the tracing with pennies.* Whose hand is bigger? How do you know that?

*Note: Instead of pennies, you can use pasta, beans, buttons, or another coin. You may want to do this activity twice using different materials to cover the hands. Talk about which materials took more or less to cover and why.

Lesson 16:   Make informal comparison of area.

© 2018 Great Minds®. eureka-math.org

Draw straight lines with your ruler to see if there are enough flowers for the butterflies.

I can draw a line to connect each butterfly with one flower.

Then I keep going to see if there are enough flowers for every butterfly.

Each butterfly gets one flower! That means there are enough!

You have 3 dog bones. Draw enough bowls so you can put 1 bone in each bowl.

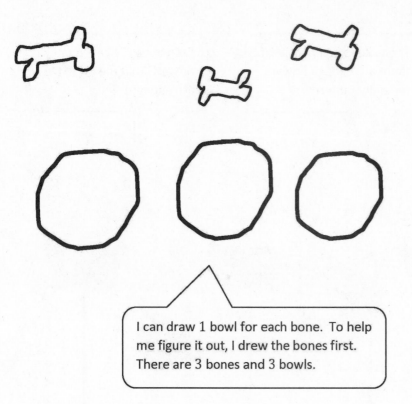

I can draw 1 bowl for each bone. To help me figure it out, I drew the bones first. There are 3 bones and 3 bowls.

Lesson 17:   Compare to find if there are enough.

Name _____   Date _____

Draw straight lines with your ruler to see if there are enough shovels for the pails.

Make sure there is a fork for every plate. Draw straight lines with a ruler from each plate to a fork. If there are not enough forks, draw one.

You have 4 fishes. Draw enough fish bowls so you can put 1 fish in each fish bowl.

Draw straight lines with your ruler to see if there are enough hats for the scarves.

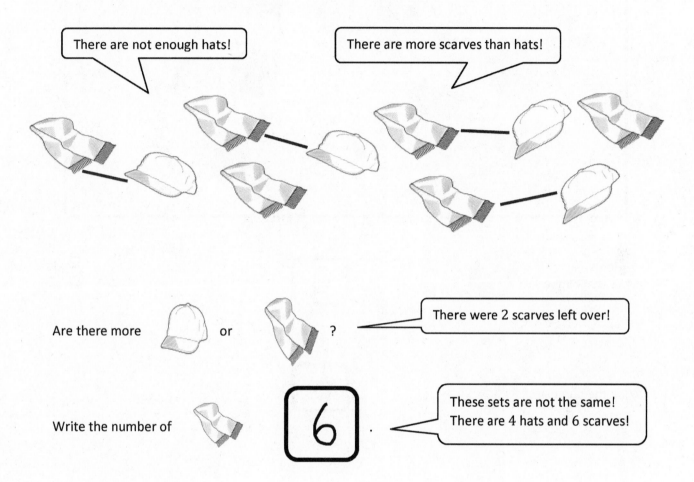

There are not enough hats!

There are more scarves than hats!

Are there more 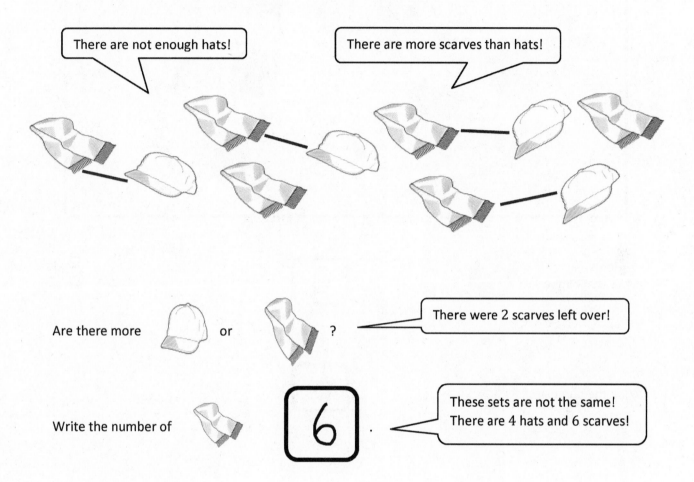 or ?

There were 2 scarves left over!

Write the number of

6.

These sets are not the same!
There are 4 hats and 6 scarves!

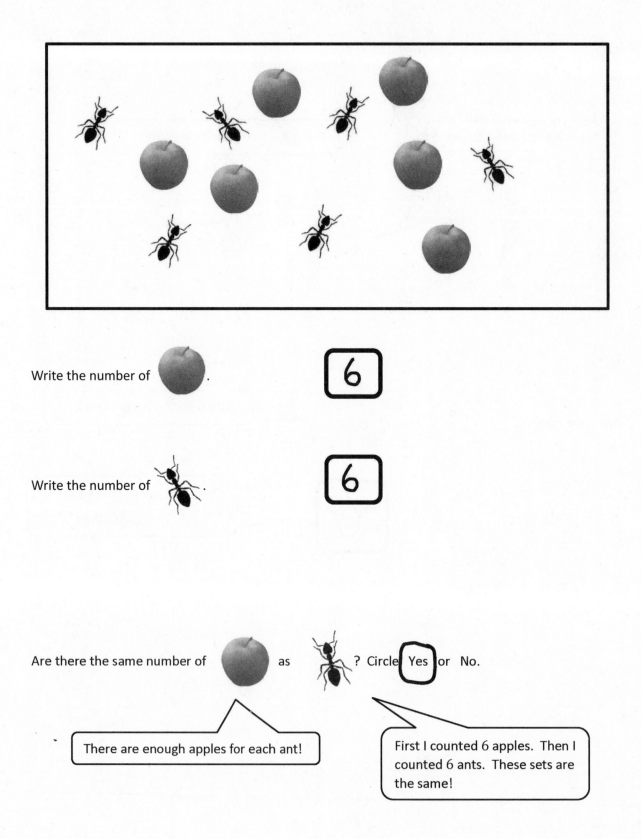

Write the number of 🍎.   6

Write the number of 🐜.   6

Are there the same number of 🍎 as 🐜? Circle Yes or No.

There are enough apples for each ant!

First I counted 6 apples. Then I counted 6 ants. These sets are the same!

Lesson 18:   Compare using *more than* and *the same as*.

EUREKA
MATH®

Name _____   Date _____

Draw straight lines with your ruler to see if there is one hoop for each ball.

Are there *more* 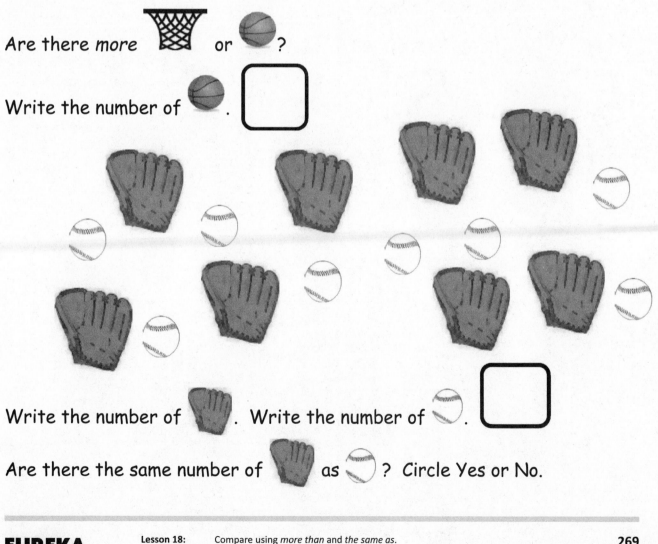 or ?

Write the number of   .

Write the number of   .   Write the number of   .

Are there the same number of   as   ?   Circle Yes or No.

Draw another bug so there are the same number of bugs as leaves.

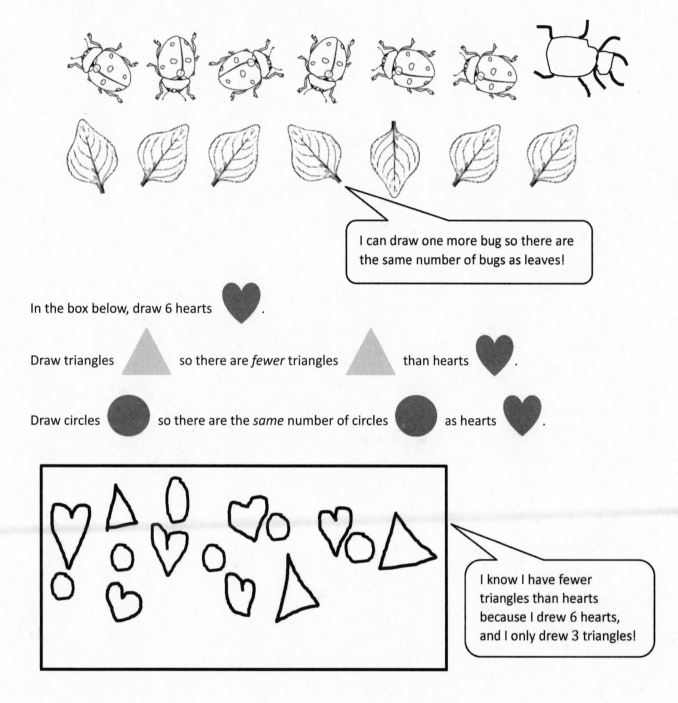

I can draw one more bug so there are the same number of bugs as leaves!

In the box below, draw 6 hearts ♥.

Draw triangles ▲ so there are *fewer* triangles ▲ than hearts ♥.

Draw circles ● so there are the *same* number of circles ● as hearts ♥.

I know I have fewer triangles than hearts because I drew 6 hearts, and I only drew 3 triangles!

Name _____   Date _____

Draw another bird so there are the same number of birds as bird cages.

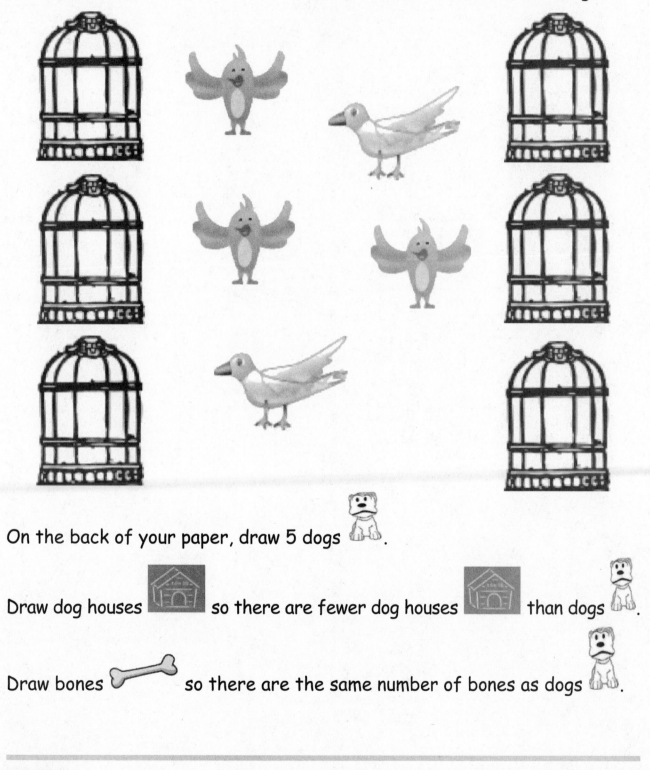

On the back of your paper, draw 5 dogs .

Draw dog houses [dog house] so there are fewer dog houses [dog house] than dogs [dog].

Draw bones [bone] so there are the same number of bones as dogs [dog].

On the first chain, color the first 4 beads orange.

On the next chain, color more than 4 beads purple.

How many beads did you color purple? Write the number in the box.

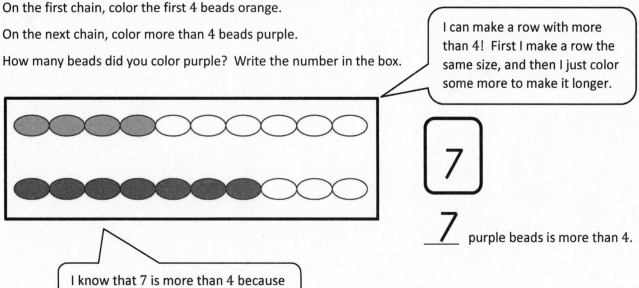

I can make a row with more than 4! First I make a row the same size, and then I just color some more to make it longer.

7

_7_ purple beads is more than 4.

I know that 7 is more than 4 because the purple row of beads is longer!

Draw a chain with more than 5 beads but fewer than 9 beads.

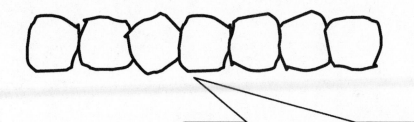

I start by making 5 beads, and then I add more, one at a time. Before I get to 9, I stop. I stopped at 7. 7 is more than 5 but still fewer than 9.

Lesson 20: Relate *more* and *less* to length.

275

Name _____   Date _____

On the first chain, color the first 3 beads blue.

On the next chain, color more than 3 beads red.

How many beads did you color red? Write the number in the box.

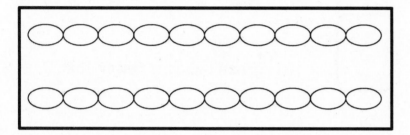

_____ red beads is more than 3.

On the first chain, color the first 5 beads green.

On the next chain, color fewer than 5 beads yellow.

How many beads did you color yellow? Write the number in the box.

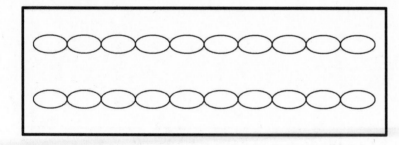

_____ yellow beads is fewer than 5.

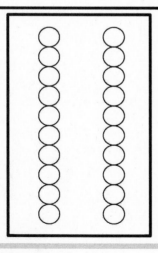

Color 2 beads brown in the first column.

Color more than 2 beads blue in the second column.

How many beads did you color in the second column?
Write the number in the box.

_____ blue beads is more than 2.

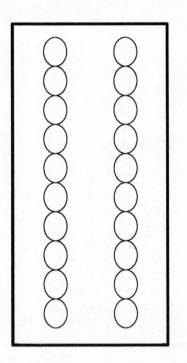

Color 9 beads red in the first column.

Color fewer than 9 beads green in the second column.

How many beads did you color in the second column? Write the number in the box.

_____ green beads is fewer than 9.

Draw a chain with more than 3 beads but fewer than 10 beads.

Draw a chain that has fewer than 10 beads but more than 4 beads.

Which has more? The 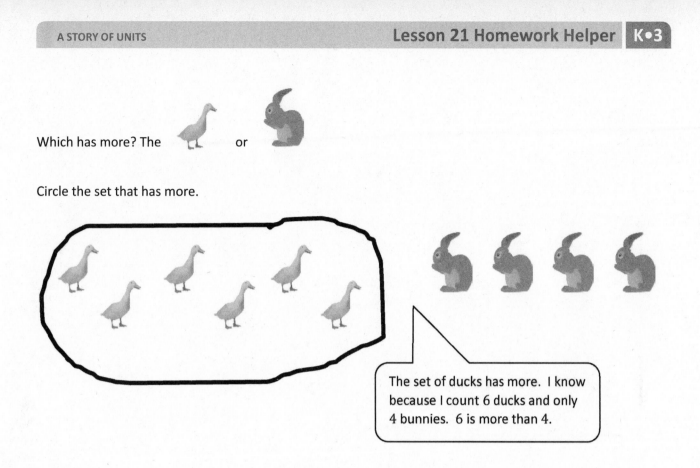 or

Circle the set that has more.

The set of ducks has more. I know because I count 6 ducks and only 4 bunnies. 6 is more than 4.

Draw a set of 3 kittens. Then draw some puppies. Are there fewer kittens or fewer puppies?

I know there are fewer puppies. I draw the 3 kittens, and then when I draw the puppies, I stop at 2.

Name _____ Date _____

Which has more? The  or ?

Circle the set that has more.

Which has fewer? The or ?

Circle the set that has fewer.

Which has fewer? The ★ or ☾ ?

Circle the set that has fewer.

On the back of your paper, draw a set of 5 books. Draw some apples. Are there fewer apples or fewer books?

Lesson 21: Compare sets informally using *more*, *less*, and *fewer*.

281

© 2018 Great Minds®. eureka-math.org

Count the fish. In the next box, draw the same number of bowls as fish.

I count 5 fish. So, I need to draw 5 bowls.

There are the same number of bowls as fish!

EUREKA MATH®

Lesson 22: Identify and create a set that has the same number of objects.

283

© 2018 Great Minds®. eureka-math.org

Name _____   Date _____

Count the birds. In the next box, draw the same number of nests as birds.

Count the houses. In the next box, draw the same number of trees as houses.

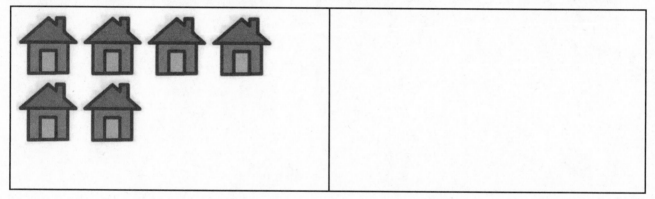

Count the monkeys. In the next box, draw the same number of bananas as monkeys.

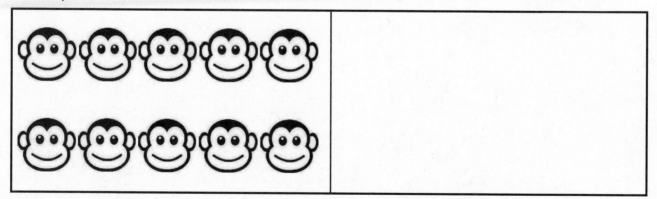

On the back of your paper, draw some pencils. Then, draw a crayon for each pencil.

There are 4 snails!

How many snails?   4

Draw a leaf for every snail and one more leaf.

How many leaves?   5

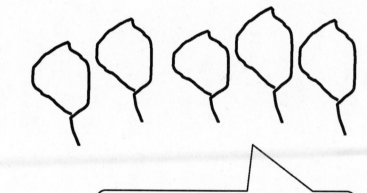

I draw 4 leaves, and then I draw 1 more. 1 more than 4 is 5.

Name _____   Date _____

How many cats? ☐

Draw a ball for every cat and 1 more ball.

How many balls? ☐

How many elephants? ☐

Draw a peanut for every elephant and 1 more peanut.

How many peanuts? ☐

Count the set of objects, and write how many in the box.

Draw a set of triangles that has 1 less, and write how many in the box. As you work, use your math words *less than*.

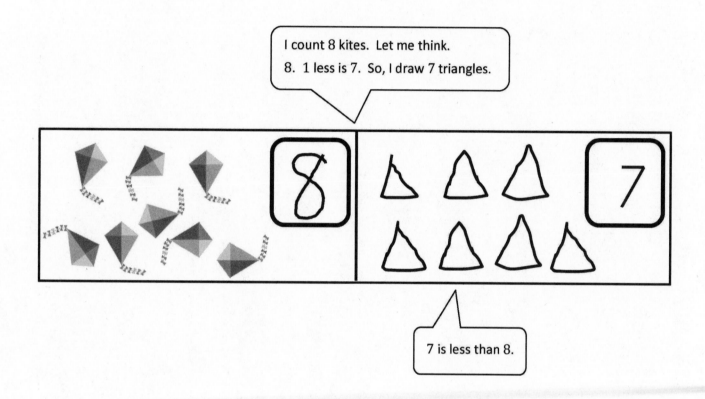

I count 8 kites. Let me think.

8. 1 less is 7. So, I draw 7 triangles.

7 is less than 8.

Lesson 24: Reason to identify and make a set that has 1 less.

291

© 2018 Great Minds®. eureka-math.org

Name _____   Date _____

Count the set of objects, and write how many in the box.

Draw a set of circles that has 1 less, and write how many in the box. As you work, use your math words *less than*.

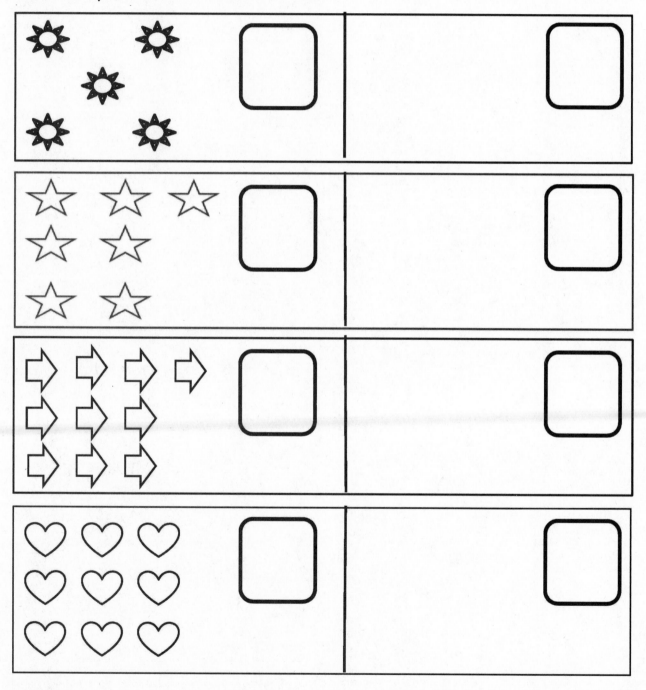

Lesson 24:   Reason to identify and make a set that has 1 less.

293

Count the objects in each line. Write how many in the box. Then, fill in the blanks below. Use the words *more than* to compare the numbers.

_8_ is more than _7_.

I can see that there is 1 more cat!
Then I counted 7 pandas and 8 cats.

8 is more than 7.

EUREKA
MATH

Lesson 25: Match and count to compare a number of objects. State which quantity is more.

295

© 2018 Great Minds®. eureka-math.org

Name _____   Date _____

Count the objects in each line. Write how many in the box. Then, fill in the blanks below.

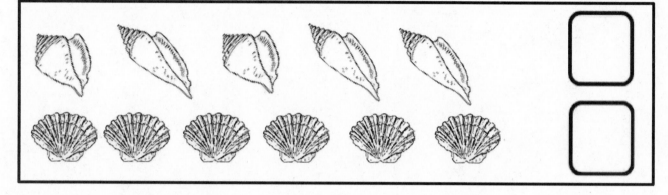

_____ is more than _____.

_____ is more than _____.

_____ is more than _____.

Lesson 25:   Match and count to compare a number of objects. State which quantity is more.

297

© 2018 Great Minds®. eureka-math.org

Count the objects in each line. Write how many in the box. Then, fill in the blanks below. Say your words *less than* out loud as you work.

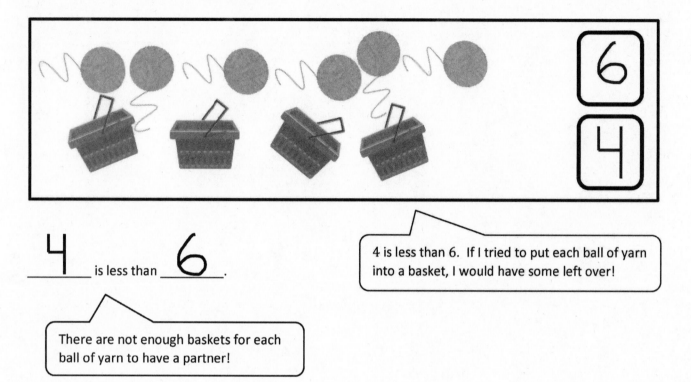

__4__ is less than __6__.

4 is less than 6. If I tried to put each ball of yarn into a basket, I would have some left over!

There are not enough baskets for each ball of yarn to have a partner!

Lesson 26: Match and count to compare two sets of objects. State which quantity is less.

299

© 2018 Great Minds®. eureka-math.org

Name _____   Date _____

Count the objects in each line. Write how many in the box. Then, fill in the blanks below.

_____ is less than _____.

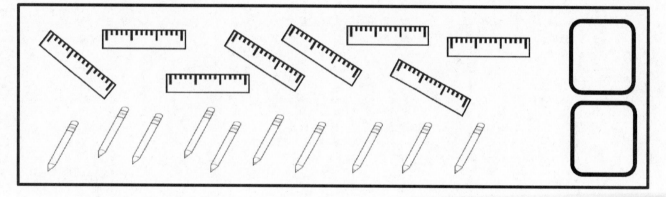

_____ is less than _____.

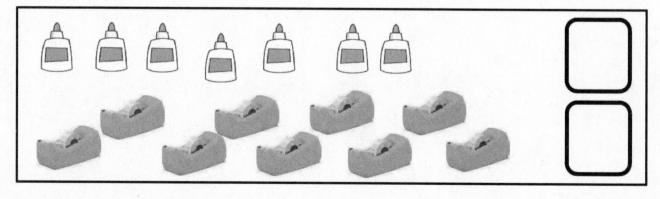

_____ is less than _____.

EUREKA MATH

Lesson 26:   Match and count to compare two sets of objects. State which quantity is less.

301

© 2018 Great Minds®. eureka-math.org

Draw a tower with more cubes.

4 is more than 3 .

3 is less than 4 .

Draw a tower with fewer cubes.

6 is more than 3 .

3 is less than 6 .

> I can make a tower with more cubes. I just make it taller! The first tower has 3 cubes, so I made a tower with 1 more. My tower has 4 cubes.

> I can make a tower with fewer cubes. I just make it shorter! The first tower has 6 cubes, so I made mine with only 3 cubes. 3 is less than 6.

Name _____ Date _____

Draw a tower with more cubes.

_____ is more than _____.

Draw a tower with fewer cubes.

_____ is more than _____.

_____ is less than _____.

Draw a train with more cubes.

_____ is more than _____.

_____ is less than _____.

On the back, draw a tower. Draw another tower that has more cubes.

_____ is more than _____. _____ is less than _____.

Visualize the number in Set A and Set B. Write the number in the sentences.

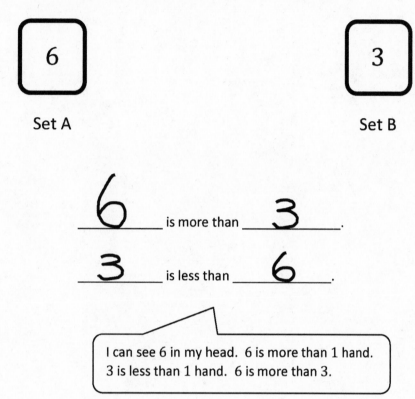

6 _____ is more than _____3_____ .

3 _____ is less than _____6_____ .

I can see 6 in my head. 6 is more than 1 hand.
3 is less than 1 hand. 6 is more than 3.

Name _____   Date _____

Visualize the number in Set A and Set B. Write the number in the sentences.

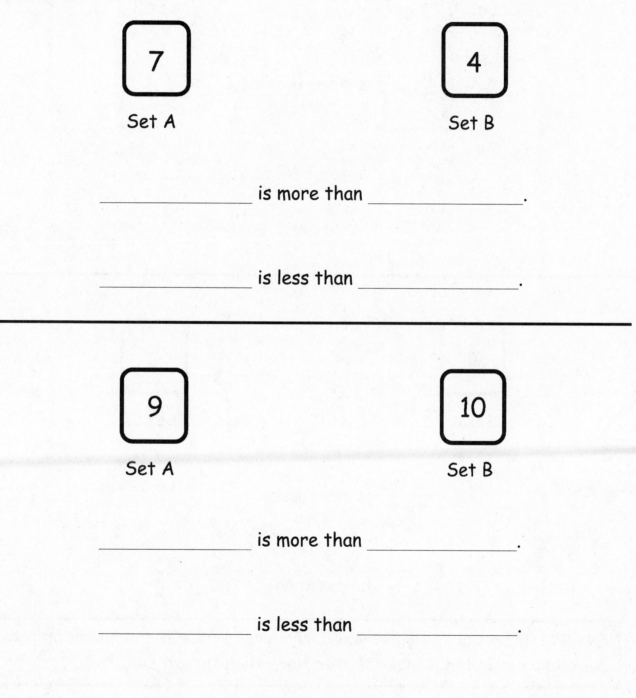

7

Set A

4

Set B

_____ is more than _____.

_____ is less than _____.

9

Set A

10

Set B

_____ is more than _____.

_____ is less than _____.

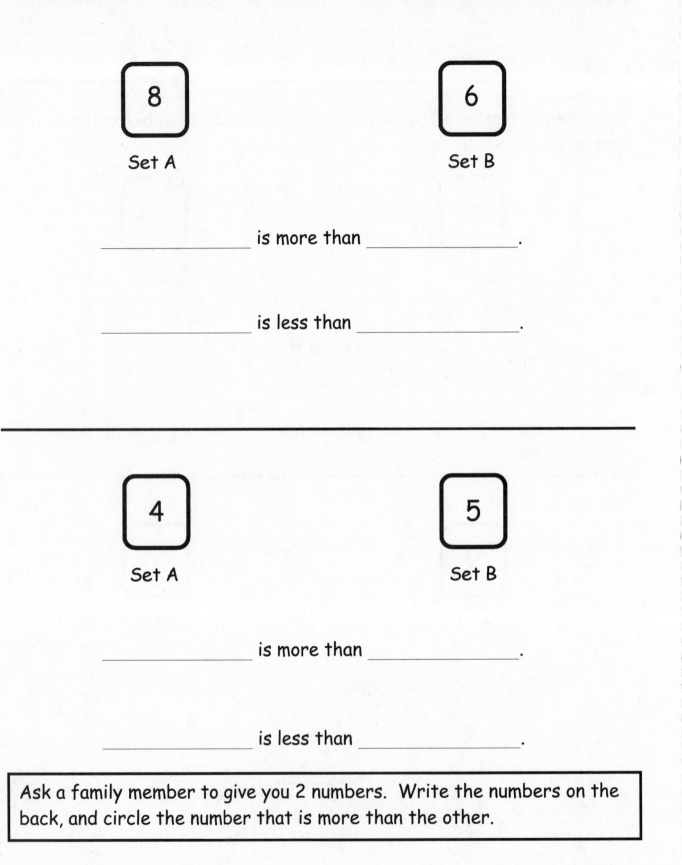

8

Set A

6

Set B

_____ is more than _____.

_____ is less than _____.

4

Set A

5

Set B

_____ is more than _____.

_____ is less than _____.

Ask a family member to give you 2 numbers. Write the numbers on the back, and circle the number that is more than the other.

EUREKA MATH

Draw a line from each container to the word that describes the amount of liquid the container is holding.

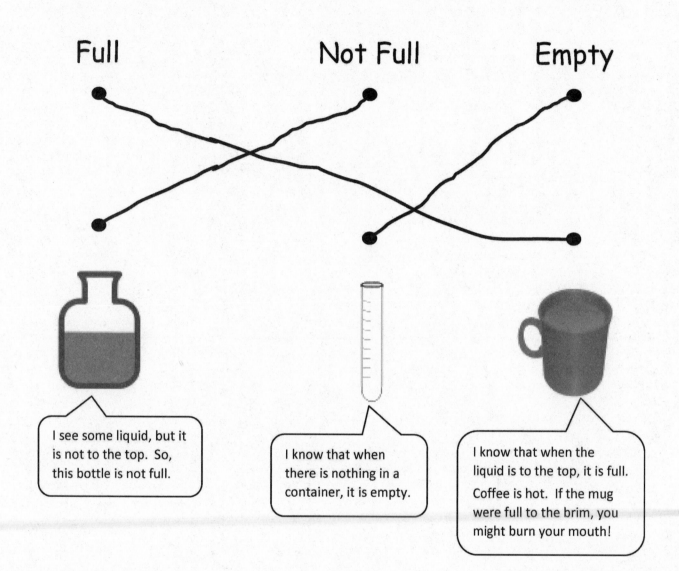

Lesson 29:   Observe cups of colored water of equal volume poured into a variety of container shapes.

© 2018 Great Minds®. eureka-math.org

Name _____ Date _____

Draw a line from each container to the word that describes the amount
of liquid the container is holding.

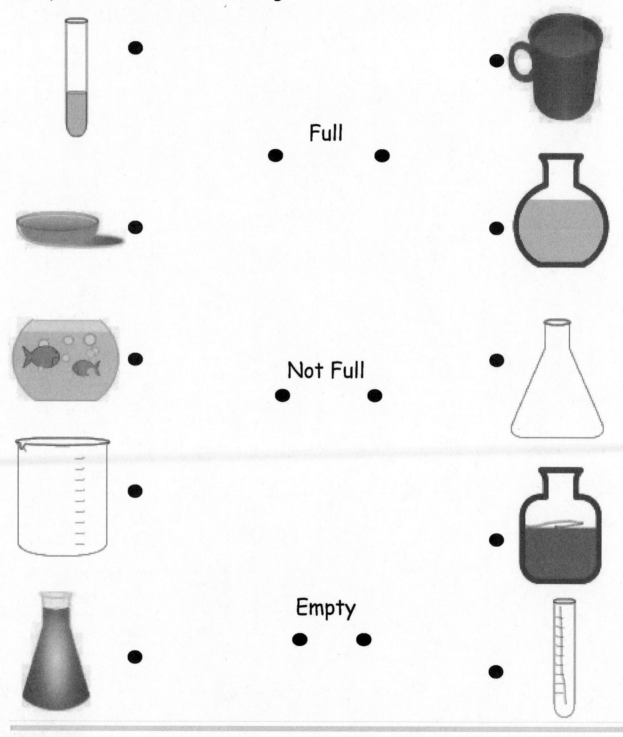

Full

Not Full

Empty

EUREKA MATH

Lesson 29:   Observe cups of colored water of equal volume poured into a variety
of container shapes.

313

Lesson 30:   Use balls of clay of equal weights to make sculptures.

Color 5 apples.

There are 10 apples in all.
I color 5 of them.
I can count the rest to see how many more to make

I colored __5__ apples.

I need to color __5__ more to make 10.

Dear Parents:

In class, we used balls of clay that weigh the same on the balance scale to make different sculptures. We saw that the same balls of clay can take various forms without changing the weight. The balls weighed the same on the balance scale, as did the sculptures.

Today's homework is a review of fluency work.

Lesson 30:   Use balls of clay of equal weights to make sculptures.

317

© 2018 Great Minds®. eureka-math.org

Name _____   Date _____

Color 4 apples.

I colored _____ apples.

I need to color _____ more to make 10.

Color 2 apples.

I colored _____ apples.

I need to color _____ more to make 10.

Color 7 apples.

I colored _____ apples.

I need to color _____ more to make 10.

Color 1 apples.

I colored _____ apples.

I need to color _____ more to make 10.

Color 9 apples.

I colored _____ apples.

I need to color _____ more to make 10.

Color 3 apples.

I colored _____ apples.

I need to color _____ more to make 10.

Lesson 30:   Use balls of clay of equal weights to make sculptures.

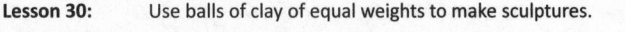

Read the following directions to your child to make a house:

- Draw a square as wide as a fork.

- Draw a triangle on top of the square as tall as your pinky for the roof.

- Draw a rectangle as long as your thumb for the door.

- Draw 2 square windows each as long as a fingernail.

EUREKA MATH®

Lesson 31:   Use benchmarks to create and compare rectangles of different lengths to make a city.

319

© 2018 Great Minds®. eureka-math.org

Name _____ Date _____

Read the following directions to your child to make a castle:

- Draw a rectangle as long as a spoon.
- Draw another rectangle on each side of the rectangle you just made.
- Draw a triangle on top of each rectangle to make towers shorter than your hand.
- Draw 1 rectangle flag pole as long as your pointer finger.
- Draw 1 square flag as long as your pinky.
- Draw a door as long as your thumb.
- Draw 2 hexagon windows each as long as a fingernail.
- Draw a prince or princess in your castle.

Castle

Lesson 31: Use benchmarks to create and compare rectangles of different lengths to make a city.

321

© 2018 Great Minds®. eureka-math.org

Circle groups of dots. Then, fill in the blanks to make a number sentence.

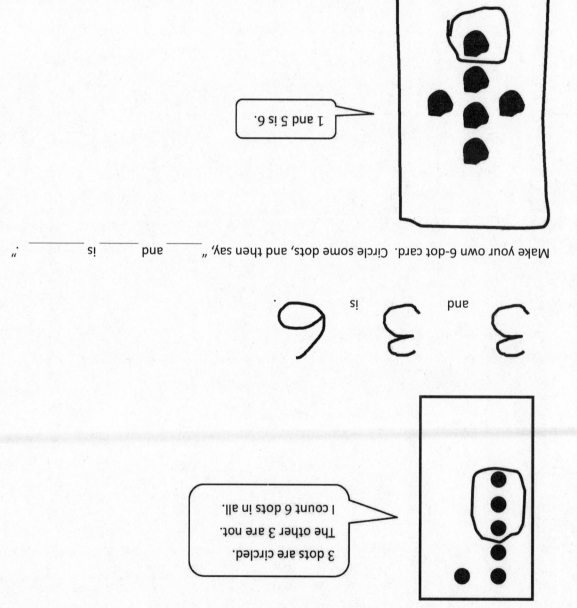

3 dots are circled.
The other 3 are not.
I count 6 dots in all.

3 and 3 is 6.

Make your own 6-dot card. Circle some dots, and then say, "_____ and _____ is _____."

1 and 5 is 6.

EUREKA MATH

Lesson 32: Culminating task—describe measurable attributes of single objects.

323

Name _____ Date _____

The homework is a review of fluency skills from Module 3.

Circle a group of dots. Then, fill in the blanks to make a number sentence.

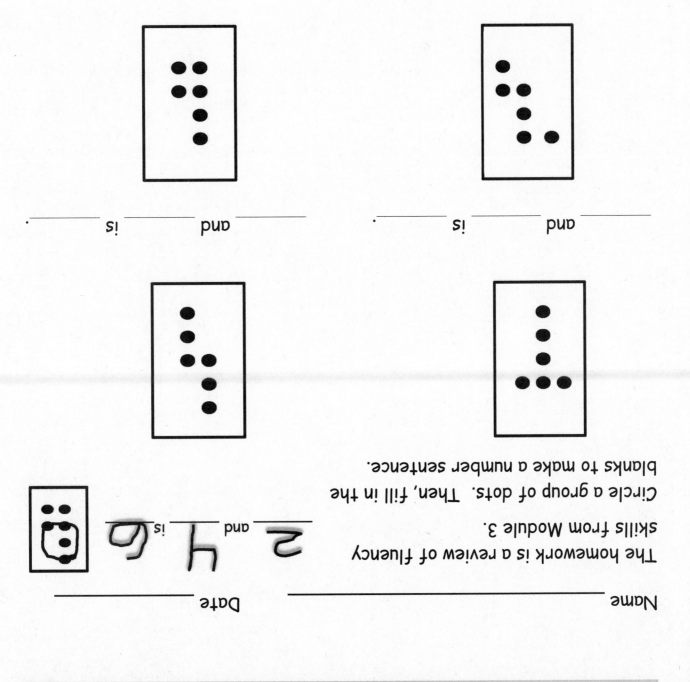

2 and 4 is 6

_____ and _____ is _____.

_____ and _____ is _____.

_____ and _____ is _____.

_____ and _____ is _____.

On the back, make your own 6-dot cards. Circle some dots, and then say, "_____ and _____ is _____."

Credits

Great Minds® has made every effort to obtain permission for the reprinting of all copyrighted material. If any owner of copyrighted material is not acknowledged herein, please contact Great Minds for proper acknowledgment in all future editions and reprints of this module.